Introduction to Colloid and Surface Chemistry

Fourth Edition

Duncan J. Shaw, BSc, Phd, FRSC
Formerly Principal Lecturer in Physical Chemistry,
Department of Chemistry and Biochemistry,
Liverpool Polytechnic

BUTTERWORTH
HEINEMANN

OXFORD AMSTERDAM BOSTON LONDON NEW YORK PARIS
SAN DIEGO SAN FRANCISCO SINGAPORE SYDNEY TOKYO

Introduction to Colloid and Surface Chemistry, 4th Edition
Duncan J. Shaw
ISBN: 9780750611824

Authorized English reprint published by Beijing World Publishing Corporation, Ltd.

《胶体和表面化学导论 第 4 版》
ISBN: 9787523223444

Notice

To Ann

Contents

Preface

This book has been written to fill a gap in the literature by offering a standard and overall coverage of colloid and surface chemistry intermediate between the brief accounts found in most textbooks of physical chemistry and the comprehensive accounts found in specialised treatises on colloid and/or surface chemistry.

In writing the book, I have kept a number of audiences in mind – particularly: university and polytechnic students studying for an honours degree or its equivalent, or commencing a programme of postgraduate research; scientists in industry who desire a broad background in a subject which may have been somewhat neglected during academic training; and those interested in branches of natural science, for whom an understanding of colloid and surface phenomena is essential.

The subject matter is, in general, approached from a fundamental angle, and the reader is assumed to possess a knowledge of the basic principles of physical chemistry. Opportunities have also been taken to describe many of the practical applications of this subject. In addition, some numerical problems (with answers) and a list of references for further reading (mainly books and review articles) are given at the end of the book.

The general character of this fourth edition is similar to that of the third edition. The text has been revised and updated throughout, the major change being the extension of Chapter 5 to include a section on the composition and structure of solid surfaces.

I wish to thank my colleagues, particularly Dr A.L. Smith, for their many helpful suggestions, and my wife, Ann, for her help in preparing the manuscript and checking the text.

D.J.S.
Southport, 1991

1 *The colloidal state*

Introduction

Colloid science concerns systems in which one or more of the components has at least one dimension within the nanometre (10^{-9}m) to micrometre (10^{-6}m) range, i.e. it concerns, in the main, systems containing large molecules and/or small particles. The adjective 'microheterogeneous' provides an appropriate description of most colloidal systems. There is, however, no sharp distinction between colloidal and non-colloidal systems.

The range of colloidal systems of practical importance is vast, as is the range of processes where colloid/surface chemical phenomena are involved.

Examples of systems which are colloidal (at least in some respects) are:

Aerosols	Foodstuffs
Agrochemicals	Ink
Cement	Paint
Cosmetics	Paper
Dyestuffs	Pharmaceuticals
Emulsions	Plastics
Fabrics	Rubber
Foams	Soil

Examples of processes which rely heavily on the application of colloid/surface phenomena are:

Adhesion	Ore flotation
Chromatography	Precipitation
Detergency	Road surfacing

Electrophoretic deposition
Emulsion polymerisation
Food processing
Grinding
Heterogeneous catalysis
Ion exchange
Lubrication
Oil-well drilling
Sewage disposal
Soil conditioning
Sugar refining
Water clarification
Water evaporation control
Water repellency
Wetting

As can be seen from the second of these lists, the existence of matter in the colloidal state may be a desirable or an undesirable state of affairs, and so it is important to know both how to make and how to destroy colloidal systems.

Colloid science is very much an interdisciplinary subject, albeit with certain areas of physics and physical chemistry most prominent. Owing to the complexity of most colloidal systems, the subject often cannot be treated readily with the exactness that tends to be associated with much of these major subject areas. It is probably a combination of this lack of precision and its interdisciplinary nature, rather than lack of importance, that has been responsible in the past for an unjustifiable tendency to neglect colloid science during undergraduate academic training.

Until the last few decades colloid science stood more or less on its own as an almost entirely descriptive subject which did not appear to fit within the general framework of physics and chemistry. The use of materials of doubtful composition, which put considerable strain on the questions of reproducibility and interpretation, was partly responsible for this state of affairs. Nowadays, the tendency is to work whenever possible with well-defined systems (e.g. monodispersed dispersions, pure surface-active agents, well-defined polymeric material) which act as models, both in their own right and for real life systems under consideration. Despite the large number of variables which are often involved, research of this nature coupled with advances in the understanding of the fundamental principles of physics and chemistry has made it possible to formulate coherent, if not always comprehensive, theories relating to many of the aspects of colloidal behaviour. Since it is important that colloid science be understood at both descriptive and theoretical levels, the study of this subject can range widely from relatively simple descriptive material to extremely complex theory.

The natural laws of physics and chemistry which describe the behaviour of matter in the massive and molecular states also, of course, apply to the colloidal state. The characteristic feature of colloid science lies in the relative importance which is attached to the various physicochemical properties of the systems being studied. As we shall see, the factors which contribute most to the overall nature of a colloidal system are:

Particle size
Particle shape and flexibility
Surface (including electrical) properties
Particle–particle interactions
Particle–solvent interactions

Classification of colloidal systems

Colloidal systems may be grouped into three general classifications:

1. *Colloidal dispersions* are thermodynamically unstable owing to their high surface free energy and are irreversible systems in the sense that they are not easily reconstituted after phase separation.
2. *True solutions of macromolecular material* (natural or synthetic) are thermodynamically stable and reversible in the sense that they are easily reconstituted after separation of solute from solvent.
3. *Association colloids* which are thermodynamically stable (see Chapter 4).

Dispersions

The particles in a colloidal dispersion are sufficiently large for definite surfaces of separation to exist between the particles and the medium in which they are dispersed. Simple colloidal dispersions are, therefore, two-phase systems. The phases are distinguished by the terms *dispersed phase* (for the phase forming the particles) and *dispersion medium* (for the medium in which the particles are distributed) – see Table 1.1. The physical nature of a dispersion depends, of course, on the respective roles of the constituent phases; for example, an oil-in-water (O/W) emulsion and a water-in-oil (W/O) emulsion could have almost the same overall composition, but their physical properties would be notably different (see Chapter 10).

Table 1.1 Types of colloidal dispersion

Dispersed phase	*Dispersion medium*	*Name*	*Examples*
Liquid	Gas	Liquid aerosol	Fog, liquid sprays
Solid	Gas	Solid aerosol	Smoke, dust
Gas	Liquid	Foam	Foam on soap solutions, fire-extinguisher foam
Liquid	Liquid	Emulsion	Milk, mayonnaise
Solid	Liquid	Sol, colloidal suspension; paste (high solid concentration)	Au sol, AgI sol; toothpaste
Gas	Solid	Solid foam	Expanded polystyrene
Liquid	Solid	Solid emulsion	Opal, pearl
Solid	Solid	Solid suspension	Pigmented plastics

Sols and emulsions are by far the most important types of colloidal dispersion. The term *sol* is used to distinguish colloidal suspensions from macroscopic suspensions; there is, of course, no sharp line of demarcation. When the dispersion medium is aqueous, the term *hydrosol* is usually used. If the dispersed phase is polymeric in nature, the dispersion is called a *latex* (pl. *latices* or *latexes*).

Foams are somewhat different in that it is the dispersion medium which has colloidal dimensions.

The importance of the interface

A characteristic feature of colloidal dispersions is the large area-to-volume ratio for the particles involved. At the interfaces between the dispersed phase and the dispersion medium characteristic surface properties, such as adsorption and electric double layer effects, are evident and play a very important part in determining the physical properties of the system as a whole. It is the material within a molecular layer or so of the interface which exerts by far the greatest influence on particle–particle and particle–dispersion medium interactions.

Despite this large area-to-volume ratio, the amount of material required to give a significant molecular coverage and modification of the interfaces in a typical colloidal dispersion can be quite small, and substantial modification of the overall bulk properties of a colloidal dispersion can often be effected by small quantities of suitable additives. For example, pronounced changes in the consistency of certain clay suspensions (such as those used in oil-well drilling) can be effected by the addition of small amounts of calcium ions (thickening) or phosphate ions (thinning)[18].

Surface science is, therefore, closely linked with colloid science; indeed, colloid science is inevitably a part of surface science, although the reverse does not necessarily hold.

The surface or interfacial phenomena associated with colloidal systems such as emulsions and foams are often studied by means of experiments on artificially prepared flat surfaces rather than on the colloidal systems themselves. Such methods provide a most useful indirect approach to the various problems involved.

Lyophilic and lyophobic systems

The terms *lyophilic* (liquid-loving) and *lyophobic* (liquid-hating) are frequently used to describe the tèndency of a surface or functional group to become wetted or solvated. If the liquid medium is aqueous, the terms *hydrophilic* and *hydrophobic* are used.

Lyophilic surfaces can be made lyophobic, and vice versa. For example, clean glass surfaces, which are hydrophilic, can be made hydrophobic by a coating of wax; conversely, the droplets in a hydrocarbon oil-in-water emulsion, which are hydrophobic, can be made hydrophilic by the addition of protein to the emulsion, the protein molecules adsorbing on to the droplet surfaces.

This terminology is particularly useful when one considers the phenomenon of surface activity. The molecules of surface-active materials have a strong affinity for interfaces, because they contain both hydrophilic and lipophilic (oil-loving) regions.

The general usage of the terms 'lyophilic' and 'lyophobic' in describing colloidal systems is somewhat illogical. 'Lyophobic' traditionally describes liquid dispersions of solid or liquid particles produced by mechanical or chemical action; however, in these so-called 'lyophobic sols' (e.g. dispersions of powdered alumina or silica in water) there is often a high affinity between the particles and the dispersion medium – i.e. the particles are really lyophilic. Indeed, if the term 'lyophobic' is taken to imply no affinity between particles and dispersion medium (an unreal situation), then the particles would not be wetted and no dispersion could, in fact, be formed. 'Lyophilic' traditionally describes soluble macromolecular material; however, lyophobic regions are often present. For example, proteins are partly hydrophobic (hydrocarbon regions) and partly hydrophilic (peptide linkages, and amino and carboxyl groups).

Structural characteristics

Experimental methods

The experimental procedures for determining particle size and shape can roughly be categorised, as follows:

1. Observation of the movement of particles in response to an applied force (see Chapter 2).
2. Direct observation of particle images (microscopy and electron microscopy) (see Chapter 3).
3. Observation of the response of particles to electromagnetic radiation (see Chapter 3).
4. Measurements which relate to the total surface area of the particles (gas adsorption and adsorption from solution) (see Chapters 5 and 6).

Particle shape

Particle asymmetry is a factor of considerable importance in determining the overall properties (especially those of a mechanical nature) of colloidal systems. Roughly speaking, colloidal particles can be classified according to shape as *corpuscular*, *laminar* or *linear* (see, for example, the electron micrographs in Figure 3.2). The exact shape may be complex but, to a first approximation, the particles can often be treated theoretically in terms of models which have relatively simple shapes (Figure 1.1).

The easiest model to treat theoretically is the sphere, and many colloidal systems do, in fact, contain spherical or nearly spherical particles. Emulsions, latexes, liquid aerosols, etc., contain spherical particles. Certain protein molecules are approximately spherical. The crystallite particles in dispersions such as gold and silver iodide sols are sufficiently symmetrical to behave like spheres.

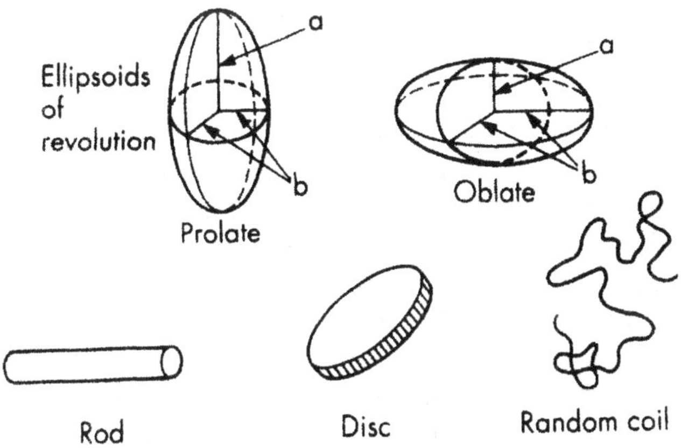

Figure 1.1 Some model representations for non-spherical particles

Corpuscular particles which deviate from spherical shape can often be treated theoretically as ellipsoids of revolution. Many proteins approximate this shape. An ellipsoid of revolution is characterised by its axial ratio, which is the ratio of the single half-axis a to the radius of revolution b. The axial ratio is greater than unity for a prolate (rugby-football-shaped) ellipsoid, and less than unity for an oblate (discus-shaped) ellipsoid.

Iron(III) oxide and clay suspensions are examples of systems containing plate-like particles.

High-polymeric material usually exists in the form of long thread-like straight or branched-chain molecules. As a result of inter-chain attraction or cross-linking (arising from covalent bonding, hydrogen bonding or van der Waals forces) and entanglement of the polymer chains, these materials often exhibit considerable mechanical strength and durability. This is not possible when the particles are corpuscular or laminar.

In nature, thread-like polymeric material fulfils an essential structural role. Plant life is built mainly from cellulose fibres. Animal life is built from linear protein material such as collagen in skin, sinew and bone, myosin in muscle and keratin in nails and hair. The coiled polypeptide chains of the so-called globular proteins which circulate in the body fluids are folded up to give corpuscular particles.

When particles aggregate together, many different shapes can be formed. These do not necessarily correspond to the shape of the primary particles.

Flexibility

Thread-like high-polymer molecules show considerable flexibility due to rotation about carbon–carbon and other bonds. In solution, the shape of these molecules alters continuously under the influence of thermal motion and a rigid rod model is therefore unsuitable. A better theoretical treatment is to consider the polymer molecules as random coils, but even this model is not completely accurate. Rotation about bonds does not permit complete flexibility, and steric and excluded volume effects also oppose the formation of a truly random configuration, so that, in these respects, dissolved linear polymer molecules will tend to be more extended than random coils. The relative magnitudes of polymer–polymer and polymer–solvent forces must also be taken into account. If the segments of the polymer chain tend to stick to one another, then a tighter than random coil, and possibly precipitation, will result; whereas a looser coil results when the polymer segments tend to avoid one another because of strong solvation and/or electrical repulsion.

Solvation

Colloidal particles are usually solvated, often to the extent of about one molecular layer, and this tightly bound solvent must be treated as a part of the particle.

Sometimes much greater amounts of solvent can be immobilised by mechanical entrapment within particle aggregates. This occurs when voluminous flocculent hydroxide precipitates are formed. In solutions of long thread-like molecules the polymer chains may cross-link, chemically or physically, and/or become mechanically entangled to such an extent that a continuous three-dimensional network is formed. If all of the solvent becomes mechanically trapped and immobilised within this network, the system as a whole takes on a solid appearance and is called a *gel*.

Polydispersity and the averages

The terms *relative molecular mass* and *particle size* can only have well-defined meanings when the system under consideration is *monodispersed* – i.e. when the molecules or particles are all alike.

Colloidal systems are generally of a *polydispersed* nature – i.e. the molecules or particles in a particular sample vary in size. By virtue of their stepwise build-up, colloidal particle and polymer molecular sizes tend to have skew distributions, as illustrated in Figure 1.2, for which the Poisson distribution often offers a good approximation. Very often, detailed determination of relative molecular mass or particle size distribution is impracticable and less perfect experimental methods, which yield average values, must be accepted. The significance of the word *average* depends on the relative contributions of the various molecules or particles to the property of the system which is being measured.

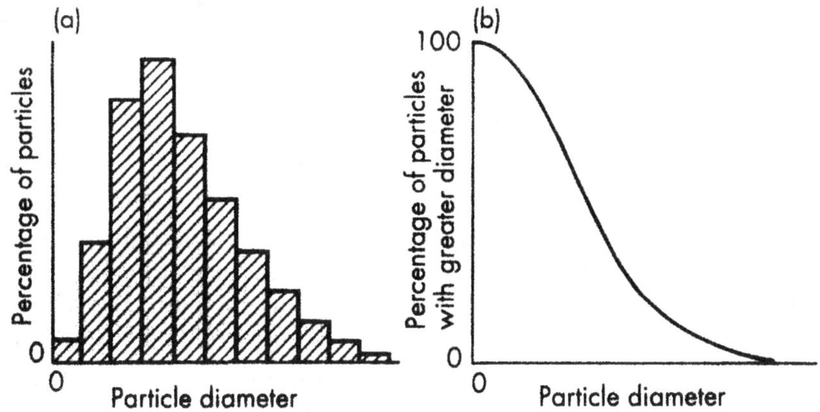

Figure 1.2 Particle diameter distribution for a polydispersed colloidal dispersion expressed (a) in histogram form, and (b) as a cumulative distribution

Osmotic pressure, which is a colligative property, depends simply on the number of solute molecules present and so yields a *number-average* relative molecular mass:

$$M_r \text{ (number average)} = \frac{\Sigma n_i M_{r,i}}{\Sigma n_i} \tag{1.1}$$

where n_i is the number of molecules of relative molecular mass $M_{r,i}$.

In most cases the larger particles make a greater individual contribution to the property being measured. If the contribution of each particle is proportional to its mass (as in light scattering), a *mass-average* relative molecular mass or particle mass is given:

$$M_r \text{ (mass average)} = \frac{\Sigma n_i M_{r,i}^2}{\Sigma n_i M_{r,i}} \tag{1.2}$$

For any polydispersed system, M_r (mass average) > M_r (number average), and only when the system is monodispersed will these averages coincide. The ratio M_r (mass average)/M_r (number average) is a measure of the degree of polydispersity.

Preparation and purification of colloidal systems

Colloidal dispersions

Basically, the formation of colloidal material involves either degradation of bulk matter or aggregation of small molecules or ions.

Dispersion of bulk material by simple grinding in a colloid mill or by ultrasonics does not, in general, lead to extensive subdivision, owing to the tendency of smaller particles to reunite (*a*) under the influence of the mechanical forces involved and (*b*) by virtue of the attractive forces between the particles. After prolonged grinding the distribution of particle sizes reaches an equilibrium. Somewhat finer dispersions can be obtained by incorporating an inert diluent to reduce the chances of the particles in question encountering one another during the grinding, or by wet-milling in the presence of surface-active material. As an example of the first of these techniques, a sulphur sol in the upper colloidal range can be prepared by grinding a mixture of sulphur and glucose, dispersing the resulting powder in water and then removing the dissolved glucose from the sol by dialysis.

A higher degree of dispersion is usually obtainable when a sol is prepared by an aggregation method. Aggregation methods involve the formation of a molecularly dispersed supersaturated solution from which the material in question precipitates in a suitably divided form. A variety of methods, such as the substitution of a poor solvent for a good one, cooling and various chemical reactions, can be utilised to achieve this end.

A coarse sulphur sol can be prepared by pouring a saturated solution of sulphur in alcohol or acetone into water just below boiling point. The alcohol or acetone vaporises, leaving the water-insoluble sulphur colloidally dispersed. This technique is convenient for dispersing wax-like material in an aqueous medium.

Examples of hydrosols which can be prepared by suitably controlled chemical reaction include the following:

1. *Silver iodide sol.* Mix equal volumes of aqueous solutions (10^{-3} to 10^{-2} mol dm^{-3}) of silver nitrate and potassium iodide. Separate the sol from larger particles by decantation or filtration. By arranging for the silver nitrate or the potassium iodide to be in very slight excess, positively or negatively charged particles, respectively, of silver iodide can be formed.
2. *Gold sol.* Add 1 cm^3 of 1% $HAuCl_4.3H_2O$ to 100 cm^3 of distilled water. Bring to the boil and add 2.5 cm^3 of 1% sodium citrate. Keep the solution just boiling. A ruby red gold sol forms after a few minutes.
3. *Sulphur sol.* Mix equal volumes of aqueous solutions (10^{-3} to 5×10^{-3} mol dm^{-3}) of $Na_2S_2O_3$ and HCl.
4. *Hydrous iron(III) oxide sol.* Add, with stirring, 2 cm^3 of 30% $FeCl_3$(aq) to 500 cm^3 of boiling distilled water. A clear reddish-brown dispersion is formed.

Nucleation and growth

The formation of a new phase during precipitation involves two distinct stages – *nucleation* (the formation of centres of crystallisation) and *crystal growth* – and (leaving aside the question of stability) it is the relative rates of these processes which determine the particle size of the precipitate so formed. A high degree of dispersion is obtained when the rate of nucleation is high and the rate of crystal growth is low.

The initial rate of nucleation depends on the degree of supersaturation which can be reached before phase separation occurs, so that colloidal sols are most easily prepared when the substance in question has a very low solubility. With material as soluble as, for example, calcium carbonate, there is a tendency for the smaller particles to redissolve (see page 68) and recrystallise on the larger particles as the precipitate is allowed to age.

The rate of particle growth depends mainly on the following factors:

1. The amount of material available.
2. The viscosity of the medium, which controls the rate of diffusion of material to the particle surface.
3. The ease with which the material is correctly orientated and incorporated into the crystal lattice of the particle.
4. Adsorption of impurities on the particle surface, which act as growth inhibitors.
5. Particle–particle aggregation.

Von Weimarn (1908) investigated the dependence on reagent concentration of the particle sizes of barium sulphate precipitates formed in alcohol–water mixtures by the reaction

$$Ba(CNS)_2 + MgSO_4 \rightarrow BaSO_4 + Mg(CNS)_2$$

At very low concentrations, *c.* 10^{-4} to 10^{-3} mol dm^{-3}, the supersaturation is sufficient for extensive nucleation to occur, but crystal growth is limited by the availability of material, with the result that a sol is formed. At moderate concentrations, *c.* 10^{-2} to 10^{-1} mol dm^{-3}, the extent of nucleation is not much greater, so that more material is available for crystal growth and a coarse filterable precipitate is formed. At very high concentrations, *c.* 2 to 3 mol dm^{-3}, the high viscosity of the medium slows down the rate of crystal growth sufficiently to allow time for much more extensive nucleation and the formation of very many small particles. Owing to their

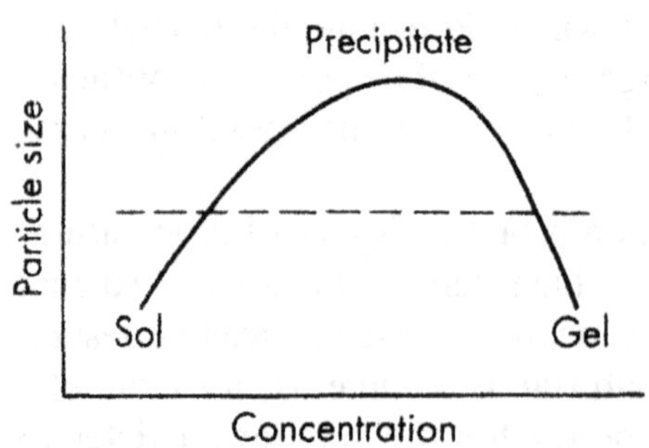

Figure 1.3 The dependence of particle size on reagent concentration for the precipitation of a sparingly soluble material

closeness, the barium sulphate particles will tend to link and the dispersion will take the form of a translucent, semi-solid gel.

The ageing of dispersions is discussed on page 68.

Monodispersed sols

Aggregation methods usually lead to the formation of polydispersed sols, mainly because the formation of new nuclei and the growth of established nuclei occur simultaneously, and so the particles finally formed are grown from nuclei formed at different times. In experiments designed to test the validity of theories, however, there are obvious advantages attached to the use of monodispersed systems. The preparation of such systems requires conditions in which nucleation is restricted to a relatively short period at the start of the sol formation. This situation can sometimes be achieved either by seeding a supersaturated solution with very small particles or under conditions which lead to a short burst of homogeneous nucleation.

An example of the seeding technique is based on that of Zsigmondy (1906) for preparing approximately monodispersed gold sols. A hot dilute aqueous solution of $HAuCl_4$ is neutralised with potassium carbonate and a part of the solute is reduced with a small amount of white phosphorus to give a highly dispersed gold sol with an average particle radius of *c.* 1 nm. The remainder of the $HAuCl_4$ is then reduced relatively slowly with formaldehyde in the presence of these small gold particles. Further nucleation is thus effectively avoided and all of the gold produced in this second stage accumulates on the seed particles. Since the absolute differences in the seed particle sizes are not great, an approximately monodispersed sol is formed. By regulating the amount of $HAuCl_4$ reduced in the second stage and the number of seed particles produced in the first stage, the gold particles can be grown to a desired size.

A similar seeding technique can be used to prepare monodispersed polymer latex dispersions by emulsion polymerisation (see page 17).

Among the monodispersed sols which have been prepared under conditions which lead to a short burst of homogeneous nucleation are (*a*) sulphur sols[132], formed by mixing very dilute aqueous solutions of HCl and $Na_2S_2O_3$; (*b*) silver bromide sols[133], by controlled cooling of hot saturated aqueous solutions of silver bromide; and (*c*) silver bromide and silver iodide sols[133], by diluting aqueous solutions of the

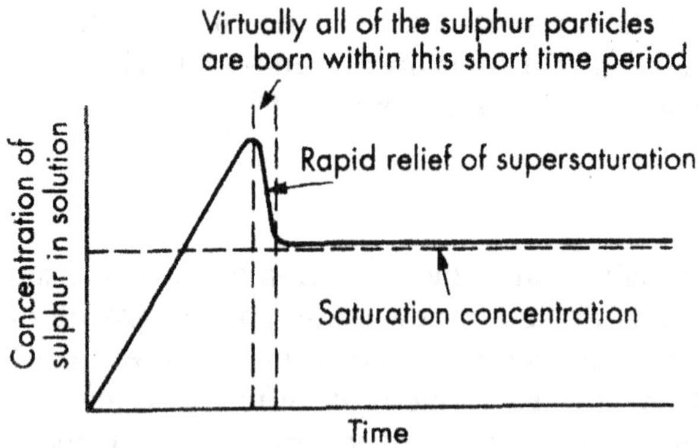

Figure 1.4 Formation of an approximately monodispersed sulphur sol by the slow reaction between $Na_2S_2O_3$ and HCl in dilute aqueous solution

complexes formed in the presence of excess silver or halide ions. In each case the concentration of the material of the dispersed phase slowly passes the saturation point and attains a degree of supersaturation at which nucleation becomes appreciable. Since the generation of dispersed phase material is slow, the appearance of nuclei and the accompanying relief of supersaturation is restricted to a relatively short period and few new nuclei are formed after this initial outburst. The nuclei then grow uniformly by a diffusion-controlled process and a sol of monodispersed particles is formed.

Various methods are also available for the preparation of monodispersed hydrous metal oxide sols[19] and silica sols[20, 134].

Monodispersed polystyrene sols are used as calibration standards for electron microscopes, light scattering photometers, Coulter counters, particle sieves, etc. Monodispersed silica is used for antireflection lens coatings. Monodispersity (even at a modest level) can usefully be exploited in photographic film, magnetic devices, pharmaceutical preparations and catalysis.

Macromolecular colloids

Macromolecular chemistry covers a particularly wide field which includes natural polymeric material, such as proteins, cellulose, gums and natural rubber; industrial derivatives of natural polymers, such as sodium carboxymethyl cellulose, rayon and vulcanised rubber; and the purely synthetic polymers, such as polythene (polyethylene), Teflon (polytetrafluoroethylene), polystyrene, Perspex (poly (methyl methacrylate)), terylene (poly (ethylene terephthalate)) and the nylons, e.g. (poly (hexamethylene adipamide)). Only brief mention of some of the more general aspects of polymerisation will be made. The reader is referred to the various specialised texts for details of preparation, properties and utilisation of these products.

High polymers contain giant molecules which are built up from a large number of similar (but not necessarily identical) units (or monomers) linked by primary valence bonds. Polymerisation reactions can be performed either in the bulk of the monomer material or in solution. A further technique, emulsion polymerisation, which permits far greater control over the reaction, is discussed on page 16.

There are two distinct types of polymerisation: addition polymerisation and condensation polymerisation.

Addition polymerisation does not involve a change of chemical composition. In general, it proceeds by a chain mechanism, a typical series of reactions being:

1. Formation of free radicals from a catalyst (initiator), such as a peroxide.
2. Initiation: for example,

$$\underset{\text{vinyl monomer}}{CH_2 = CHX} + \underset{\text{free radical}}{\dot{R}} \rightarrow RCH_2 - \dot{C}HX$$

3. Propagation:

$$RCH_2 - \dot{C}HX + CH_2 = CHX \rightarrow RCH_2 - CHX - CH_2 - \dot{C}HX, \text{ etc.}$$

$$\text{to } \underset{\text{vinyl polymer}}{R(CH_2 - CHX)_n\, CH_2} - \dot{C}HX$$

4. Termination. This can take place in several ways, such as reaction of the activated chain with an impurity, an additive or other activated chains, or by disproportionation between two activated chains.

A rise in temperature increases the rates of initiation and termination, so that the rate of polymerisation is increased but the average chain length of the polymer is reduced. The chain length is also reduced by increasing the catalyst concentration, since this causes chain initiation to take place at many more points throughout the reaction mixture.

Condensation polymerisation involves chemical reactions between functional groups with the elimination of a small molecule, usually water. For example,

$$\underset{\text{hexamethylenediamine}}{x\mathrm{NH_2(CH_2)_6NH_2}} + \underset{\text{adipic acid}}{x\mathrm{COOH(CH_2)_4COOH}} \rightarrow$$

$$\underset{\text{nylon 66}}{\mathrm{H[NH(CH_2)_6NHCO(CH_2)_4CO]_xOH}} + (2x-1)\mathrm{H_2O}$$

If the monomers are bifunctional, as in the above example, then a linear polymer is formed. Terminating monofunctional groups will reduce the average degree of polymerisation. Polyfunctional monomers, such as glycerol and phthalic acid, are able to form branching points, which readily leads to irreversible network formation (see Chapter 9). Bakelite, a condensation product of phenol and formaldehyde, is an example of such a space-network polymer. Linear polymers are usually soluble in suitable solvents and are thermoplastic – i.e. they can be softened by heat without decomposition. In contrast, highly condensed network polymers are usually hard, are almost completely insoluble and thermoset – i.e. they cannot be softened by heat without decomposition.

Emulsion polymerisation and polymer latexes

A polymerisation method which is of particular interest to the colloid scientist is that of emulsion polymerisation.

In bulk polymerisation, processing difficulties are usually encountered unless the degree of polymerisation is sharply limited. These difficulties arise mainly from the exothermic nature of polymerisation reactions and the necessity for efficient cooling to avoid the undesirable effects associated with a high reaction temperature (see page 15). Even at moderate degrees of polymerisation the resulting high viscosity of the reaction mixture makes stirring and efficient heat transfer very difficult.

The difficulties associated with heat transfer can be overcome, and higher molecular weight polymers obtained, by the use of an emulsion system. The heat of polymerisation is readily dissipated into the aqueous phase and the viscosity of the system changes only slightly during the reaction.

A typical recipe for the polymerisation of a vinyl monomer would be to form an oil-in-water emulsion from:

monomer (e.g. styrene)	25–50 g
emulsifying agent (e.g. fatty acid soap)	2–4 g
initiator (e.g. potassium persulphate)	0.5–1 g
chain transfer agent (e.g. dodecyl mercaptan)	0–0.2 g
water	200 g

Nitrogen is bubbled through the emulsion, which is maintained at *c.* 50–60°C for *c.* 4–6 h. The chain transfer agent limits the relative molecular mass of the polymer to *c.* 10^4, compared with *c.* 10^5–10^6 without it. The latex so formed is then purified by prolonged dialysis.

The mechanism of emulsion polymerisation is complex. The basic theory is that originally proposed by Harkins[21]. Monomer is distributed throughout the emulsion system (*a*) as stabilised emulsion droplets, (*b*) dissolved to a small extent in the aqueous phase and (*c*) solubilised in soap micelles (see page 89). The micellar environment appears to be the most favourable for the initiation of polymerisation. The emulsion droplets of monomer appear to act mainly as reservoirs to supply material to the polymerisation sites by diffusion through the aqueous phase. As the micelles grow, they adsorb free emulsifier from solution, and eventually from the surface of the emulsion droplets. The emulsifier thus serves to stabilise the polymer particles. This theory accounts for the observation that the rate of polymerisation and the number of polymer particles finally produced depend largely on the emulsifier concentration, and that the number of polymer particles may far exceed the number of monomer droplets initially present.

Monodispersed sols containing spherical polymer particles (e.g. polystyrene latexes[22–24, 135]) can be prepared by emulsion polymerisation, and are particularly useful as model systems for studying various aspects of colloidal behaviour. The seed sol is prepared with the emulsifier concentration well above the critical micelle concentration; then, with the emulsifier concentration below the critical micelle concentration, subsequent growth of the seed particles is achieved without the formation of further new particles.

Dialysis and gel filtration

Conventional filter papers retain only particles with diameters in excess of at least 1 μm and are, therefore, permeable to colloidal particles.

The use of membranes for separating particles of colloidal dimensions is termed *dialysis*. The most commonly used membranes are prepared from regenerated cellulose products such as collodion (a partially evaporated solution of cellulose nitrate in alcohol plus ether), Cellophane and Visking. Membranes with various, approximately known, pore sizes can be obtained commercially (usually in the form of 'sausage skins' or 'thimbles'). However, particle size and pore size cannot be properly correlated, since the permeability of a membrane is also affected by factors such as electrical repulsion when the membrane and particles are of like charge, and particle adsorption on the filter which can lead to a blocking of the pores.

Dialysis is particularly useful for removing small dissolved molecules from colloidal solutions or dispersions – e.g. extraneous electrolyte such as KNO_3 from AgI sol. The process is hastened by stirring so as to maintain a high concentration gradient of diffusible molecules across the membrane and by renewing the outer liquid from time to time (Figure 1.5).

Ultrafiltration is the application of pressure or suction to force the solvent and small particles across a membrane while the larger particles are retained. The membrane is normally supported between fine wire screens or deposited in a highly porous support such as a sintered glass disc. An important application of ultrafiltration is the so-called reverse osmosis method of water desalination[25].

Another most valuable development of the ultrafiltration principle is the technique of gel permeation chromatography for the separation of the components of a polymeric sample and determination of the relative molecular mass distribution. The usual experimental arrangement involves the application of a pressure to force polymer solution through a chromatographic column filled with porous beads. The larger polymer molecules tend not to enter the pores of the beads and so pass through the column relatively quickly, whereas the smaller polymer molecules tend to diffuse through the pore structure of the beads and so take longer to pass through the column. The eluted polymer can be detected and estimated by measuring the refractive

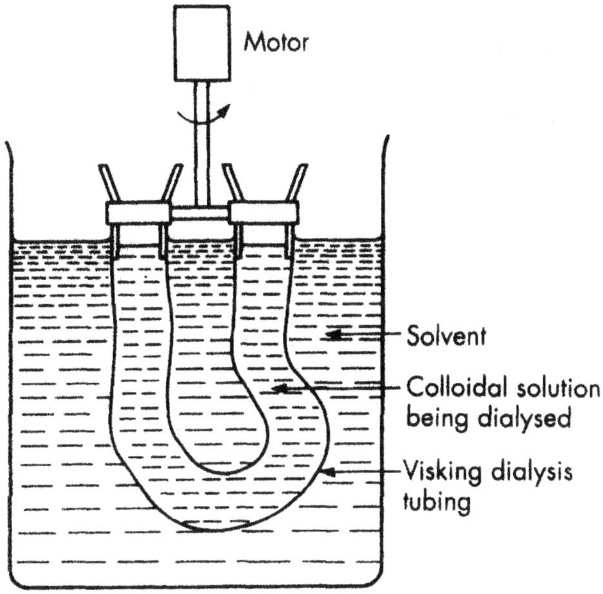

Figure 1.5 A simple dialysis set-up

index of the emerging solution, and the relationship between retention time and relative molecular mass is determined by calibrating the apparatus with polymer fractions which have been characterised by other methods, such as osmotic pressure (see page 37), light scattering (see page 57) or viscosity (see page 251).

A further modification of dialysis is the technique of electrodialysis, as illustrated in Figure 1.6. The applied potential between the metal

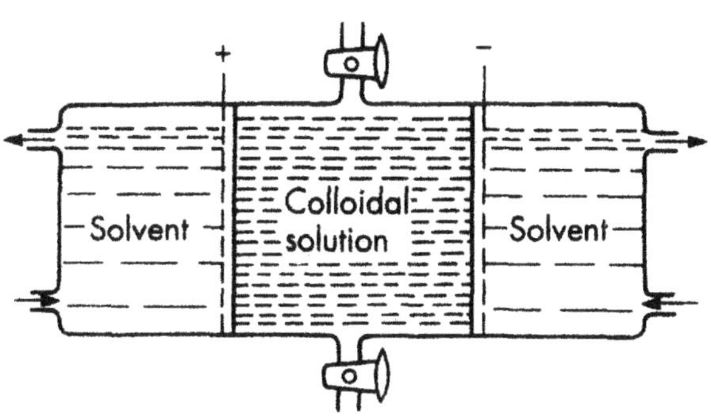

Figure 1.6 Electrodialysis

screens supporting the membranes speeds up the migration of small ions to the membrane surface prior to their diffusion to the outer liquid. The accompanying concentration of charged colloidal particles at one side and, if they sediment significantly, at the bottom of the middle compartment is termed *electrodecantation*.

2 Kinetic Properties

The motion of particles in liquid media

In this chapter the thermal motion of dissolved macromolecules and dispersed colloidal particles will be considered, as will their motion under the influence of gravitational and centrifugal fields. Thermal motion manifests itself on the microscopic scale in the form of Brownian motion, and on the macroscopic scale in the forms of diffusion and osmosis. Gravity (or a centrifugal field) provides the driving force in sedimentation. Among the techniques for determining molecular or particle size and shape are those which involve the measurement of these simple properties.

The motion of colloidal particles in an electric field is treated separately in Chapter 7.

Before these kinetic properties are discussed in any detail, some general comments on the laws governing the motion of particles through liquids are appropriate.

Sedimentation rate

Consider the sedimentation of an uncharged particle of mass m and specific volume v in a liquid of density ρ. The driving (or sedimenting) force on the particle, which is independent of particle shape or solvation, is $m\ (1 - v\rho)g$, where g is the local acceleration due to gravity (or a centrifugal field). The factor $(1 - v\rho)$ allows for the buoyancy of the liquid. The liquid medium offers a resistance to the motion of the particle which increases with increasing velocity. Provided that the velocity is not too great, which is always the case for colloidal (and somewhat larger) particles, the resistance of the liquid is, to a first approximation, proportional to the velocity of the sedimenting particle. In a very short time, a terminal velocity, $\mathrm{d}x/\mathrm{d}t$, is attained, when the driving force on the particle and the resistance of the liquid are equal:

$$m(1 - v\rho)g = f \qquad (2.1)$$

where f is the frictional coefficient of the particle in the given medium.

For spherical particles the frictional coefficient is given by Stokes' law

$$f = 6\pi\eta a \qquad (2.2)$$

where η is the viscosity of the medium and a the radius of the particle.

Therefore, if ρ_2 is the density of a spherical particle (in the dissolved or dispersed state (i.e. $\rho_2 = 1/v$)), then

$$\tfrac{4}{3}\pi a^3(\rho_2 - \rho)g = 6\pi\eta a \frac{dx}{dt}$$

or

$$\frac{dx}{dt} = \frac{2a^2(\rho_2 - \rho)g}{9\eta} \qquad (2.3)$$

The derivation of Stokes' law assumes that:

1. The motion of the spherical particle is extremely slow.
2. The liquid medium extends an infinite distance from the particle – i.e. the solution or suspension is extremely dilute.
3. The liquid medium is continuous compared with the dimensions of the particle. This assumption is valid for the motion of colloidal particles, but not for that of small molecules or ions which are comparable in size with the molecules constituting the liquid medium.

For spherical colloidal particles undergoing sedimentation, diffusion or electrophoresis, deviations from Stokes' law usually amount to much less than 1 per cent and can be neglected.

Frictional ratios

The frictional coefficient of an asymmetric particle depends on its orientation. At low velocities such particles are in a state of random orientation through accidental disturbances, and the resistance of the liquid to their motion can be expressed in terms of a frictional coefficient averaged over all possible orientations. For particles of equal volume the frictional coefficient increases with increasing asymmetry. This is because, although the resistance of the liquid is reduced when the asymmetric particle is end-on to the direction of flow, it is increased to a greater extent with side-on orientations, so that on average there is an increase in resistance. The frictional coefficient is also increased by particle solvation.

A particle containing a given volume of dry material will have its smallest possible frictional coefficient, f_0, in a particular liquid when it is in the form of an unsolvated sphere. The *frictional ratio*, f/f_0 (i.e. the ratio of the actual frictional coefficient to the frictional coefficient of the equivalent unsolvated sphere) is, therefore, a measure of a combination of asymmetry and solvation.

With application to dissolved proteins in mind, Oncley[26] has computed frictional ratios for ellipsoids of revolution of varying degrees of asymmetry and hydration. The resulting contour diagram (Figure 2.1) shows the combinations of axial ratio and hydration which are compatible with given frictional ratios. The separate contributions of asymmetry and hydration cannot be determined unless other relevant information is available.

Brownian motion and translational diffusion

Brownian motion

A fundamental consequence of the kinetic theory is that, in the absence of external forces, all suspended particles. regardless of their size, have the same average translational kinetic energy. The average translational kinetic energy for any particle is $\frac{3}{2}kT$, or $\frac{1}{2}kT$ along a given axis – i.e. $\frac{1}{2}m(\mathrm{d}x/\mathrm{d}t)^2 = \frac{1}{2}kT$, etc.; in other words, the average particle velocity increases with decreasing particle mass.

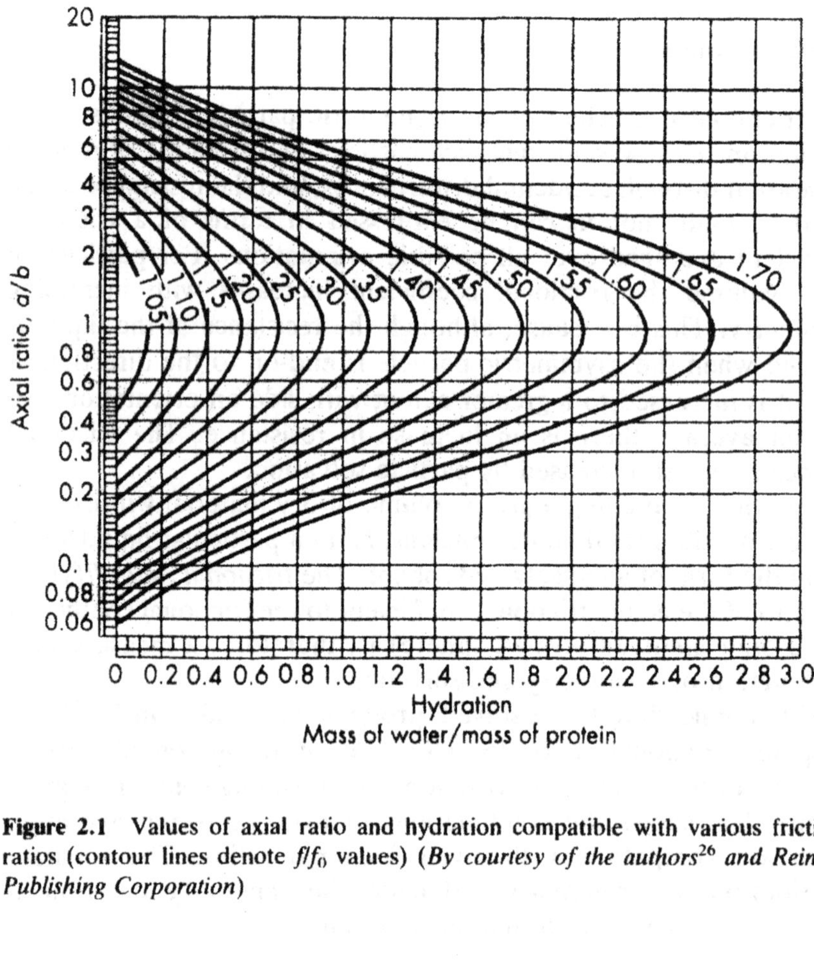

Figure 2.1 Values of axial ratio and hydration compatible with various frictional ratios (contour lines denote f/f_0 values) (*By courtesy of the authors*[26] *and Reinhold Publishing Corporation*)

The motion of individual particles is continually changing direction as a result of random collisions with the molecules of the suspending medium, other particles and the walls of the containing vessel. Each particle pursues a complicated and irregular zig-zag path. When the particles are large enough for observation, this random motion is referred to as Brownian motion, after the botanist who first observed this phenomenon with pollen grains suspended in water. The smaller the particles, the more evident is their Brownian motion.

Treating Brownian motion as a three-dimensional 'random walk', the mean Brownian displacement $\bar{x}$ of a particle from its original position along a given axis after a time t is given by Einstein's equation:

$$\bar{x} = (2Dt)^{1/2} \tag{2.4}$$

where D is the diffusion coefficient (see page 26–7).

The theory of random motion helps towards understanding the behaviour of linear high polymers in solution. The various segments of a flexible linear polymer molecule are subjected to independent thermal agitation, and so the molecule as a whole will take up a continually changing and somewhat random configuration (see page 8). The average distance between the ends of a completely flexible and random chain made up of n segments each of length l is equal to $l(n)^{1/2}$ (cf. Einstein's equation above). This average end-to-end distance becomes $l(2n)^{1/2}$ if an angle of 109° 28′ (the tetrahedral angle) between adjacent segments is specified.

The diffusion coefficient of a suspended material is related to the frictional coefficient of the particles by Einstein's law of diffusion:

$$Df = kT \tag{2.5}$$

Therefore, for spherical particles,

$$D = \frac{kT}{6\pi\rho a} = \frac{RT}{6\pi\rho a N_A} \tag{2.6}$$

where N_A is Avogadro's constant, and

$$\bar{x} = \left(\frac{RTt}{3\pi\eta a N_A}\right)^{1/2} \tag{2.7}$$

Perrin (1908) studied the Brownian displacement (and sedimentation equilibrium under gravity; see page 35) for fractionated mastic and gamboge suspensions of known particle size, and calculated values for Avogadro's constant varying between $5.5 \times 10^{23}\ \text{mol}^{-1}$ and $8 \times 10^{23}\ \text{mol}^{-1}$. Subsequent experiments of this nature have

Table 2.1 Diffusion coefficients and Brownian displacements calculated for uncharged spheres in water at 20°C

Radius	$D_{20°C}/m^2s^{-1}$	$\bar{x}$ *after* 1 h
10^{-9} m (1 nm)	2.1×10^{-10}	1.23×10^{-3} m (1.23 mm)
10^{-8} m (10 nm)	2.1×10^{-11}	3.90×10^{-4} m (390 μm)
10^{-7} m (100 nm)	2.1×10^{-12}	1.23×10^{-4} m (123 μm)
10^{-6} (1 μm)	2.1×10^{-13}	3.90×10^{-5} m (39 μm)

yielded values of N_A closer to the accepted 6.02×10^{23} mol^{-1}; for example, Svedberg (1911) calculated $N_A = 6.09 \times 10^{23}$ mol^{-1} from observations on monodispersed gold sols of known particle size in the ultramicroscope. The correct determination of Avogadro's constant from observations on Brownian motion provides striking evidence in favour of the kinetic theory

As a result of Brownian motion, continual fluctuations of concentration take place on a molecular or small-particle scale. For this reason, the second law of thermodynamics is only valid on the macroscopic scale.

Translational diffusion

Diffusion is the tendency for molecules to migrate from a region of high concentration to a region of lower concentration and is a direct result of Brownian motion.

Fick's first law of diffusion (analogous with the equation of heat conduction) states that the mass of substance dm diffusing in the x direction in a time dt across an area A is proportional to the concentration gradient dc/dx at the plane in question:

$$dm = -DA\frac{dc}{dx}dt \tag{2.8}$$

(The minus sign denotes that diffusion takes place in the direction of decreasing concentration.)

The rate of change of concentration at any given point is given by an exactly equivalent expression, Fick's second law:

$$\frac{dc}{dt} = D\frac{d^2c}{dx^2} \tag{2.9}$$

The proportionality factor D is called the diffusion coefficient. It is not strictly a constant, since it is slightly concentration-dependent.

Equations (2.4) and (2.5) can be derived using equation (2.8), as follows.

Brownian displacement equation (2.4)

Consider a plane AB (Figure 2.2) passing through a dispersion and separating regions of concentration c_1 and c_2, where $c_1 > c_2$. Let the average Brownian displacement of a given particle perpendicular to AB be $\bar{x}$ in time t. For each particle, this displacement has equal probability of being 'left to right' or 'right to left'. The net mass of particles displaced from left to right across unit area of AB in time t is, therefore given by

$$m = \frac{(c_1 - c_2)\bar{x}}{2} = \frac{(c_1 - c_2)\bar{x}^2}{2\bar{x}}$$

If $\bar{x}$ is small, $\dfrac{c_1 - c_2}{\bar{x}} = -\dfrac{\mathrm{d}c}{\mathrm{d}x}$

Therefore,

$$m = -\frac{1}{2}\frac{\mathrm{d}c}{\mathrm{d}x}\bar{x}^2 \tag{2.10}$$

From equation (2.8)

$$m = -D\frac{\mathrm{d}c}{\mathrm{d}x}t \tag{2.11}$$

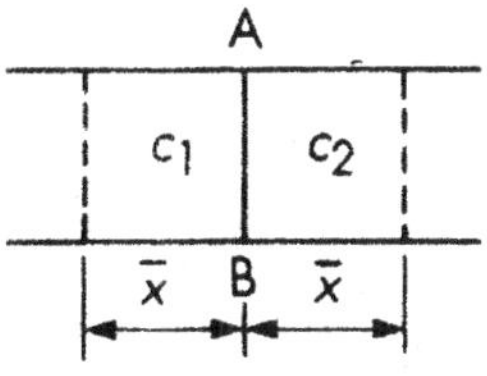

Figure 2.2

Therefore, combining equations (2.10) and (2.11),

$$\bar{x} = (2Dt)^{1/2} \tag{2.4}$$

Diffusion equation (2.5)

The work done in moving a particle through a distance dx against a frictional resistance to motion $f(\mathrm{d}x/\mathrm{d}t)$ can be equated with the resulting change in chemical potential given by the expression

$$\mathrm{d}\mu = kT\,\mathrm{d}\ln c$$

i.e. $$f\frac{\mathrm{d}x}{\mathrm{d}t}\mathrm{d}x = kT\,\mathrm{d}\ln c$$

Therefore,

$$\frac{\mathrm{d}x}{\mathrm{d}t} = \frac{kT}{f}\frac{\mathrm{d}\ln c}{\mathrm{d}x} = \frac{kT}{fc}\frac{\mathrm{d}c}{\mathrm{d}x} \tag{2.12}$$

Since

$$-\frac{\mathrm{d}m}{\mathrm{d}t} = Ac\frac{\mathrm{d}x}{\mathrm{d}t}$$

then combining this expression with equation (2.8) gives

$$c\frac{\mathrm{d}x}{\mathrm{d}t} = D\frac{\mathrm{d}c}{\mathrm{d}x} \tag{2.13}$$

Therefore, combining equations (2.12) and (2.13),

$$Df = kT \tag{2.5}$$

For a system containing spherical particles, $D = RT/6\pi\eta a N_A$ – i.e. $D \propto 1/m^{1/3}$, where m is the particle mass. For systems containing asymmetric particles, D is correspondingly smaller (see Table 2.3). Since $D = kT/f$, the ratio D/D_0 (where D is the experimental diffusion coefficient and D_0 is the diffusion coefficient of a system containing the equivalent unsolvated spheres) is equal to the reciprocal of the frictional ratio f/f_0. Charge effects are discussed on page 37.

Measurement of diffusion coefficients[27]

Free boundary methods

To study free diffusion, a sharp boundary must first be formed between the solution and the solvent (or solution of lower concentration) in a suitable diffusion cell. One of the most satisfactory techniques for achieving this end is the sliding method illustrated in Figure 2.3. The boundary can be displaced away from the ground-glass flanges to facilitate optical observation, and can be sharpened even further by gently sucking away any mixed layers with an extremely fine capillary tube inserted from above.

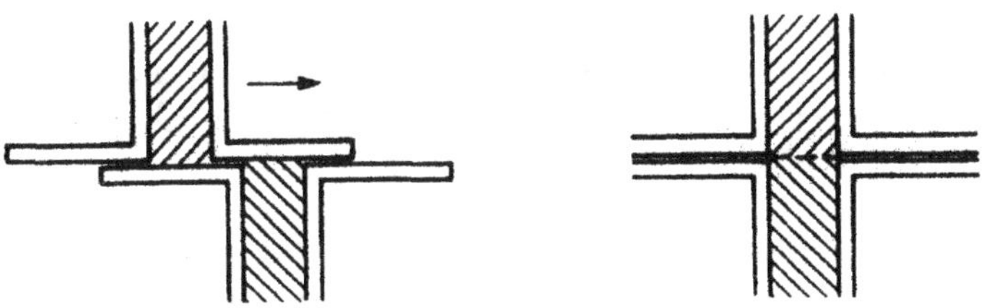

Figure 2.3 Formation of an initially sharp boundary between two miscible liquids

As diffusion proceeds, concentration and concentration gradient changes will take place as illustrated in Figure 2.4. To ensure that the broadening of the boundary is due to diffusion only, very accurate temperature control (to avoid convection currents) and freedom from mechanical vibration must be maintained. The avoidance of convection is a problem common to all kinetic methods of investigating colloidal systems.

Concentration changes are observed optically from time to time either by light absorption (e.g. ultraviolet absorption for protein solutions) or, more usually, by schlieren or interference methods. These optical methods for examining concentration changes in liquid columns (especially the schlieren technique in its various forms) are also employed in the ultracentrifuge and for studying moving-boundary electrophoresis. The schlieren method is based on the fact that, at a boundary between two transparent liquids of different

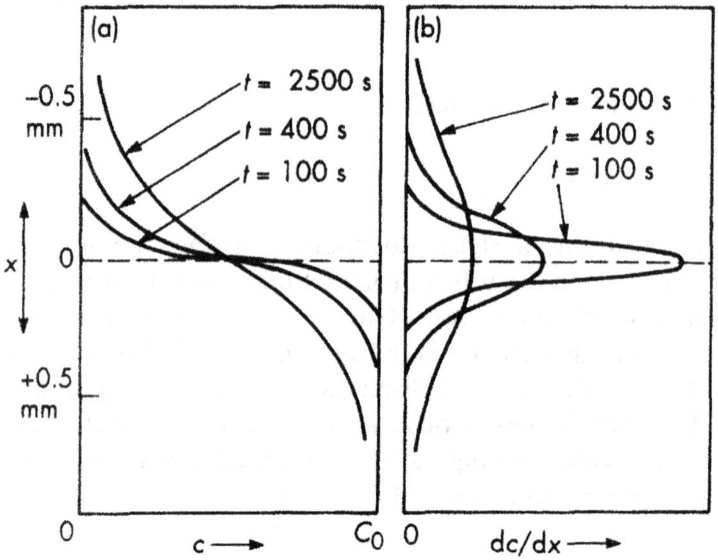

Figure 2.4 Distribution of (a) concentration and (b) concentration gradient at different times after the formation of a sharp boundary (calculated for $D = 2.5 \times 10^{-11}$ $m^2\ s^{-1}$)[28] *(By courtesy of Oxford University Press)*

refractive indexes, a beam of light perpendicular to the liquid column is refracted, thus casting a shadow which marks the region of changing refractive index. The optical system can be arranged so that the boundary is photographed in the form of a refractive index gradient peak. Since refractive index increments and concentration increments are normally proportional, the shape of the concentration gradient peak is also recorded directly.

Free diffusion columns are arranged to be sufficiently long for the initial concentrations at the extreme ends of the cell to remain unaltered during the course of the experiment. For a monodispersed system under these conditions the concentration gradient curves (Figure 2.4b) can be shown, by solving Fick's equations, to take the shape of Gaussian distribution curves represented by the expression

$$\frac{dc}{dx} = \frac{-c_0}{(4\pi Dt)^{1/2}} \exp[-x^2/4Dt] \tag{2.14}$$

from which D can be calculated. This Gaussian shape is not very sensitive to polydispersity, so that average diffusion coefficients can usually be calculated without too much difficulty.

Porous plug method

The two liquids are separated by a sintered glass disc with a pore size of 5–15 μm (the more concentrated solution on top) and kept stirred. The liquid in the pores of the disc is effectively immobilised and freed from the influence of external disturbances, so that the transport of solute through the disc is due solely to diffusion. The extent of diffusion can be determined by any analytical method.

By analogy with Fick's first law,

$$\frac{dm}{dt} = \frac{-AD(c_1 - c_2)}{l} \tag{2.15}$$

where A is the cross-section of pores, and l is the effective length of pores.

The ratio A/l is determined by calibrating the apparatus with a substance of known diffusion coefficient.

This method has considerable advantages over the free boundary methods with regard to experimental procedure. Possible objections to the method are: (*a*) the calibration of the cell with material of different relative molecular mass and/or shape from the material under investigation is not necessarily valid; and (*b*) entrapment of air bubbles in the pores or adsorption of the diffusing molecules on the pore walls will invalidate the results.

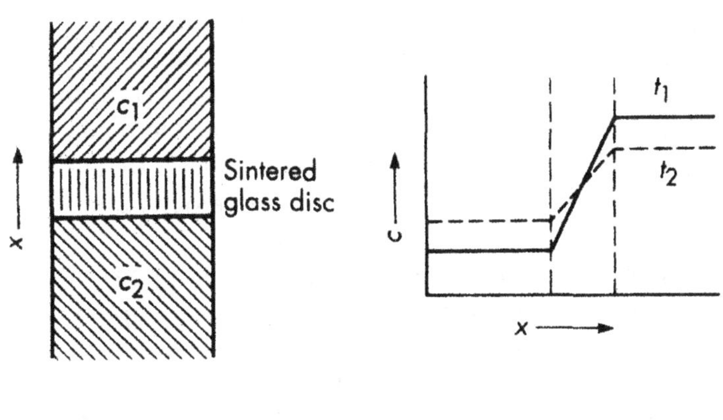

Figure 2.5 Porous plug method

Dynamic light scattering (see page 61)

The ultracentrifuge

There are a number of experimental techniques (for example, the Wiegner tube and the Odén balance)[29] which make use of sedimentation under gravity for fractionating or determining particle-size distributions in systems that contain relatively coarse suspended material such as soils and pigments. A popular method of carrying out a cumulative sedimentation size analysis is the balance method, in which the weight of particles settling out on to a balance pan is recorded as a function of time. Sedimentation under gravity has a practical lower limit of *c.* 1 μm. Smaller (colloidal) particles sediment

Table 2.2 Sedimentation rates under gravity for uncharged spheres of density 2 g cm^{-3} in water at 20°C, calculated from Stokes' law

Radius	*Sedimentation rate*
10^{-9} m (1 nm)	2.2×10^{-12} m s^{-1} (8 nm h^{-1})
10^{-8} m (10 nm)	2.2×10^{-10} m s^{-1} (0.8 μm h^{-1})
10^{-7} m (100 nm)	2.2×10^{-8} m s^{-1} (80 μm h^{-1})
10^{-6} m (1 μm)	2.2×10^{-6} m s^{-1} (8 mm h^{-1})
10^{-5} m (10 μm)	2.2×10^{-4} m s^{-1} (0.8 m h^{-1})

so slowly under gravity that the effect is obliterated by the mixing tendencies of diffusion and convection.

By employing centrifugal forces instead of gravity, the application of sedimentation can be extended to the study of colloidal systems, and has been used, in particular, for the characterisation of substances of biological origin, such as proteins, nucleic acids and viruses[30–31]. The driving force on a suspended molecule or particle then becomes $m(1-v\rho)\omega^2 x$, where ω is the angular velocity and x the distance of the particle from the axis of rotation.

An ultracentrifuge is a high-speed centrifuge equipped with a suitable optical system (usually schlieren or interference optics, the latter being particularly useful when low concentrations are involved) for recording sedimentation behaviour and with facilities for eliminating the disturbing effects of convection currents and vibration. The sample is contained in a sector-shaped cell mounted in a rotor (usually *c.*

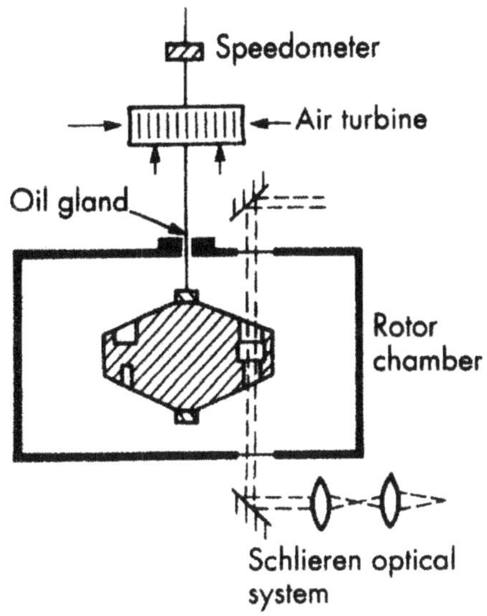

Figure 2.6 Essential features of an air-driven ultracentrifuge

18 cm diameter); this spins in a thermostatted chamber containing hydrogen at a reduced pressure. Several mechanisms for driving the rotor have been investigated – Svedberg, who pioneered this field, employed an oil turbine; these have been superseded by simpler and less expensive air-driven and electrically driven instruments.

The ultracentrifuge can be used in two distinct ways for investigating suspended colloidal material. In the velocity method a high centrifugal field (up to *c.* 400 000 *g*) is applied and the displacement of the boundary set up by sedimentation of the colloidal molecules or particles is measured from time to time (Figure 2.7). In the equilibrium method the colloidal solution is subjected to a much lower centrifugal field, until sedimentation and diffusion (mixing) tendencies balance one another and an equilibrium distribution of particles throughout the sample is attained.

Sedimentation velocity

Equating the driving force on a macromolecule in a centrifugal field with the frictional resistance of the suspending medium,

$$m(1 - v\rho)\omega^2 x = f\,\frac{dx}{dt}$$

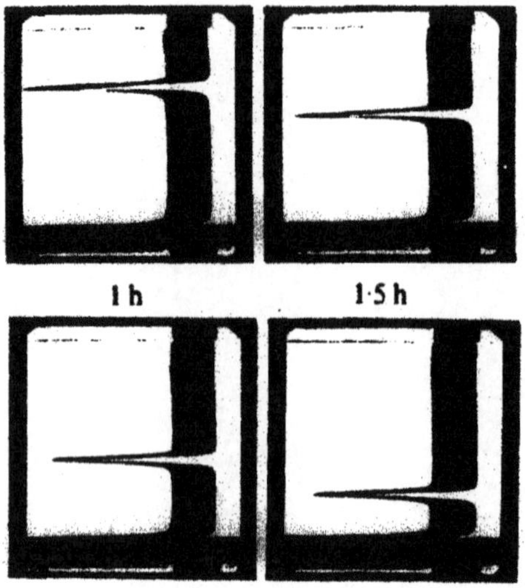

Figure 2.7 Sedimentation of a monodispersed sample of *Limulus* haemocyanin measured by the Philpot–Svenson schlieren method (18 000 rev min^{-1})[32] (*By courtesy of the American Chemical Society*)

and, since $Df=kT$,

$$\frac{dx}{dt} = \frac{mD(1-v\rho)\omega^2 x}{kT}$$

$$= \frac{MD(1-v\rho)\omega^2 x}{RT}$$

or $$M = \frac{RTs}{D(1-v\rho)} \qquad (2.16)$$

where M is the molar mass (unsolvated) and s the sedimentation coefficient

$$s = \frac{dx/dt}{\omega^2 x}$$

Integrating,

$$s = \frac{\ln x_2/x_1}{\omega^2(t_2 - t_1)} \qquad (2.17)$$

where x_1 and x_2 are the distances of the boundary from the axis of rotation at times t_1 and t_2. Therefore,

$$M = \frac{RT \ln x_2 / x_1}{D(1 - v\rho)(t_2 - t_1)\omega^2} \tag{2.18}$$

It is evident from the above expressions that the appropriate diffusion coefficient must also be measured in order that molecular or particle masses may be determined from sedimentation velocity data. In this respect, a separate experiment is required, since the diffusion coefficient cannot be determined accurately *in situ*, because there is a certain self-sharpening of the peak due to the sedimentation coefficient increasing with decreasing concentration.

Care must be taken to ensure that the system under investigation remains uncoagulated. This applies to any technique for determining molecular or particle masses. s, D and v are corrected to a standard temperature, usually 20°C, and should be extrapolated to zero concentration.

With polydispersed systems either a broadening of the boundary (in addition to that caused by diffusion) or the formation of distinct peaks representing the various fractions is observed. Sedimentation does not provide an unequivocal method for establishing the homogeneity of a colloidal system. For example, a mixture of serum albumin and haemoglobin is homogeneous with respect to sedimentation velocity but the two proteins are easily distinguished from each other by electrophoresis.

Knowledge of M and v enables D_0 and, hence, the ratio D_0/D (the frictional ratio) to be calculated.

Sedimentation equilibrium

Consider the flow of molecules or particles across an area A in a colloidal solution where the concentration is c and the concentration gradient is $\mathrm{d}c/\mathrm{d}x$. The rate of flow is $cA\,(\mathrm{d}x/\mathrm{d}t)$ due to sedimentation, and, from Fick's first law, $-DA\,(\mathrm{d}c/\mathrm{d}x)$ due to diffusion. When sedimentation equilibrium is attained, the net flow is zero, so that

$$c\frac{\mathrm{d}x}{\mathrm{d}t} = D\frac{\mathrm{d}c}{\mathrm{d}x}$$

and, since

$$\frac{dx}{dt} = \frac{MD(1 - v\rho)\omega^2 x}{RT}$$

then

$$\frac{dc}{c} = \frac{\omega^2 M(1 - v\rho)x\,dx}{RT}$$

Integrating,

$$M = \frac{2RT \ln c_2 / c_1}{\omega^2 (1 - v\rho)(x_2^2 - x_1^2)} \qquad (2.19)$$

where c_1 and c_2 are the sedimentation equilibrium concentrations at distances x_1 and x_2 from the axis of rotation. Ideal behaviour has been assumed.

When a state of sedimentation–diffusion equilibrium has been reached, the molecular or particle mass can, therefore, be evaluated without a knowledge of the diffusion coefficient (and, hence, independently of shape and solvation) by determining relative concentrations at various distances from the axis of rotation. Molecules as small as sugars have been studied by this technique.

Polydispersity introduces complications and is reflected by a drift of M with x. Conversely, consistency of M with x is an indication of sample homogeneity with respect to M.

The disadvantage of the sedimentation equilibrium technique is that the establishment of equilibrium may take as long as several days, which not only is inconvenient generally, but also accentuates the importance of avoiding convectional disturbances.

From a theoretical analysis of the intermediate stages during which the solute is being redistributed, Archibald[136] has developed a technique which involves measurements at intervals during the early stages of the sedimentation equilibrium experiment and so does not entail a long wait for equilibrium to be established. The ratio s/D can be calculated from the expression

$$\frac{dc / dx'}{c'x'} = \frac{\omega^2 s}{D} \qquad (2.20)$$

where dc/dx' is the concentration gradient at the meniscus, c' the concentration at the meniscus and x' the distance of the meniscus from the axis of rotation.

Charge effects

The treatment of sedimentation and diffusion is a little more complicated when the particles under consideration are charged. The smaller counter-ions (see Chapter 7) tend to sediment at a slower rate and lag behind the sedimenting colloidal particles. A potential is thus set up which tends to restore the original condition of overall electrical neutrality by accelerating the motion of the counter-ions and retarding the motion of the colloidal particles.

The reverse situation applies to diffusion. The smaller counter-ions tend to diffuse faster than the colloidal particles and drag the particles along with them and increase their rate of diffusion.

These effects can be overcome by employing swamping electrolyte concentrations. Any potentials which might develop are then readily dissipated by a very small displacement of a large number of counter-ions.

Osmotic pressure

The measurement of a colligative property (i.e. lowering of vapour pressure, depression of freezing point, elevation of boiling point or osmotic pressure) is a standard procedure for determining the relative molecular mass of a dissolved substance. Of these properties, osmotic pressure is the only one with a practical value in the study of macromolecules. Consider, for example, a solution of 1 g of macromolecular material of relative molecular mass 50 000 dissolved in 100 cm^3 of water. Assuming ideal behaviour, the depression of freezing point would be 0.0004 K and the osmotic pressure (at 25°C) would be 500 Nm^{-2} (i.e. 5 cm water). This freezing point depression would be far too small to be measured with sufficient accuracy by conventional methods and, even more important, it would be far too sensitive to small amounts of low relative molecular mass impurity; in fact, it would be doubled by the presence of just 1 mg of impurity of relative molecular mass 50. Not only does osmotic pressure provide a measurable effect, but also the effect of any low relative molecular mass material to which the membrane is permeable can virtually be eliminated.

The usefulness of osmotic pressure measurements is, nevertheless, limited to a relative molecular mass range of about 10^4–10^6. Below 10^4, permeability of the membrane to the molecules under consideration might prove to be troublesome; and above 10^6, the osmotic pressure will be too small to permit sufficiently accurate measurements.

Osmosis takes place when a solution and a solvent (or two solutions of different concentrations) are separated from each other by a semipermeable membrane – i.e. a membrane which is permeable to the solvent but not to the solute. The tendency to equalise chemical potentials (and, hence, concentrations) on either side of the membrane results in a net diffusion of solvent across the membrane. The counter-pressure necessary to balance this osmotic flow is termed the *osmotic pressure*.

Osmosis can also take place in gels and constitutes an important swelling mechanism.

The osmotic pressure Π of a solution is described in general terms by the so-called viral equation

$$\Pi = cRT\left(\frac{1}{M} + B_2c + B_3c^2 + \ldots\right) \tag{2.21}$$

where c is the concentration of the solution (expressed as mass of solute divided by volume of solution), M is the molar mass of the solute, and B_2, B_3, etc., are constants.

Therefore,

$$M = RT / \lim_{c\to 0} \Pi / c \tag{2.22}$$

Deviations from ideal behaviour are relatively small for solutions of compact macromolecules such as proteins but can be quite appreciable for solutions of linear polymers. Such deviations have been treated thermodynamically[137,138], mainly in terms of the entropy change on mixing, which is considerably greater (especially for linear polymers dissolved in good solvents) than the ideal entropy change on mixing for a system obeying Raoult's law. This leads to solvent activities which are smaller than ideal – i.e. an apparent increase in the concentration and an actual increase in the osmotic pressure of the polymer solution.

The resulting relative molecular mass refers to the composition of

Table 2.3 Molecular data of proteins and other substances in aqueous solution

Name	$\frac{s_{20°C}}{10^{-13}\ s}$	$\frac{D_{20°C}}{10^{-11}\ m^2\ s^{-1}}$	$\frac{v_{20°C}}{cm^3\ g^{-1}}$	M_r (*s*)	M_r (*e*)	M_r (π)	f/f_0	*Isoelectric point* (pH)*
Urea		129				60		
Sucrose		36				342		
Ribonuclease	1.85	13.6	0.709	12 700	13 000		1.04	
Myoglobin	2.04	11.3	0.741	16 900	17 500	17 000	1.11	7.0
Gliadin	2.1	6.7	0.724	27 500	27 000		1.6	
β-Lactoglobulin	3.1	7.3	0.751	41 000	38 000	35 000	1.26	5.2
Ovalbumin	3.55	7.8	0.749	44 000	40 500	45 000	1.16	4.55
Haemoglobin (horse)	4.48	6.3	0.749	68 000	68 000	67 000	1.24	6.9
Serum albumin (horse)	4.46	6.1	0.748	70 000	68 000	73 000	1.27	4.8
Serum globulin (horse)	7.1	4.0	0.745	167 000	150 000	175 000	1.4	
Fibrinogen (bovine)	8.2	2.0	0.706	330 000			2.3	5.2
Myosin	7.2	0.8	0.74	840 000			4.0	5.4
Bushy stunt virus	132	1.15	0.739	10 600 000			1.27	4.1
Tobacco mosaic virus	174	0.3	0.727	59 000 000			2.9	

M_r (*s*) = relative molecular mass from sedimentation velocity measurements
M_r (*e*) = relative molecular mass from sedimentation equilibrium measurements
M_r (π) = relative molecular mass from osmotic pressure measurements
*measured at 20°C in acetate or phosphate buffer at an ionic strength of 0.02 mol kg^{-1}

the solute with respect to solvation, etc., which was used in establishing the concentration of the solution. For polydispersed systems a number average is measured.

Measurement of osmotic pressure

A great deal of work has been devoted to the preparation of suitable semipermeable membranes and to perfecting sensitive methods for measuring osmotic pressure[33,34].

Certain practical difficulties arise if the solution is simply allowed to rise and seek its own equilibrium level:

1. If the liquid rises up a narrow tube and the total volume of solution is large, the liquid level may be too sensitive to temperature fluctuations during the course of the experiment to permit reliable measurements. In addition, a capillary rise correction must be made under these conditions. To overcome the difficulty associated with a sticking meniscus, which is often encountered in the study of aqueous solutions, a liquid of low surface tension and good wetting properties, such as toluene or petrol ether, is used in the capillary.

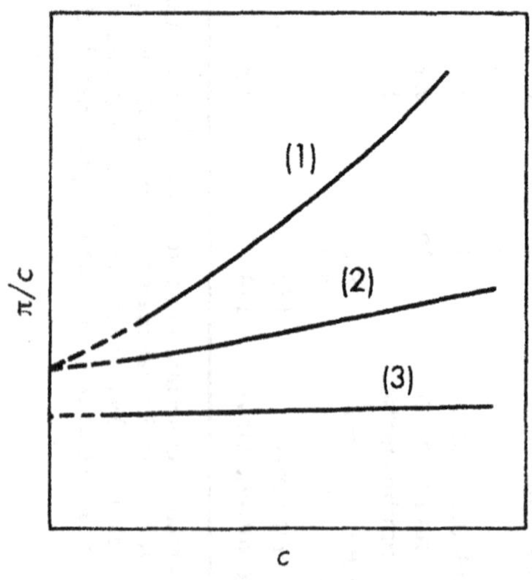

Figure 2.8 Dependence of reduced osmotic pressure on concentration: (1) a linear high polymer in a good solvent; (2) the same polymer in a poor solvent; (3) a globular protein in aqueous solution

2. If the liquid rises up a wide tube and the volume of solution is small, significant changes in concentration will take place and the establishment of equilibrium will be extremely slow.

A frequently adopted procedure for overcoming these difficulties is to set the liquid level in turn at slightly above and slightly below the anticipated equilibrium level and then plot its position as a function of time. If the estimation of the final equilibrium level has been reasonably accurate, the two curves will be almost symmetrical, and by plotting the half-sum as a function of time the equilibrium level can be reliably estimated after relatively short times (Figure 2.9). This procedure permits the use of a moderately wide capillary tube; however, good thermostatting and a firmly supported membrane are still required.

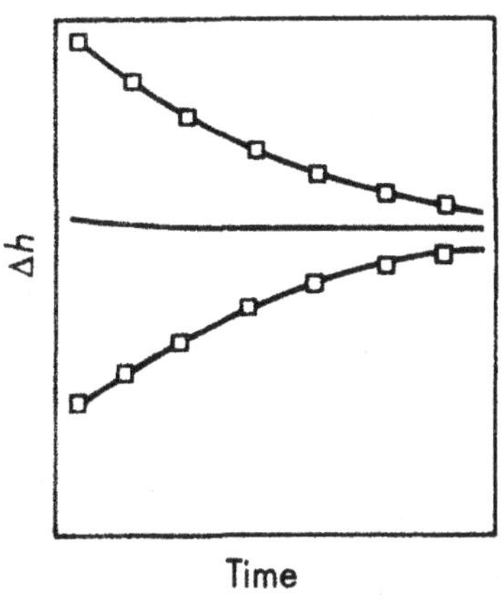

Figure 2.9 Estimation of osmotic pressure by the half-sum method

In the Fuoss–Mead osmometer[139] the membrane is firmly clamped (and also acts as a gasket) between two carefully machined stainless steel blocks provided with channels into which small volumes of solvent and solution are introduced (Figure 2.10). Owing to the high ratio of membrane surface area to solution volume (*c.* 75 cm^2: 15 cm^3), the approach to equilibrium is rapid. With last membranes, and using the half-sum method, measurements in well under 1 h are possible. Over such a short time the membrane can be slightly permeable to the macromolecules without introducing serious errors.

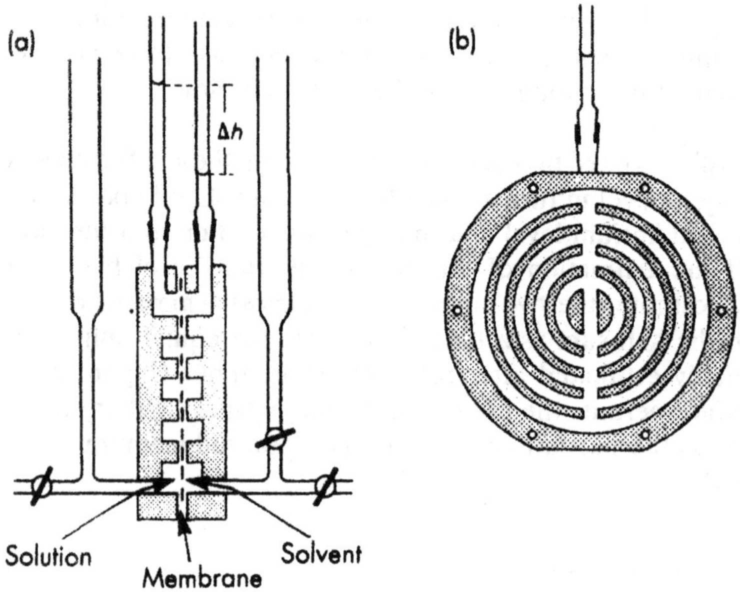

Figure 2.10 Schematic representation of the Fuoss–Mead osmometer: (a) vertical cross-section; (b) inner surface of each half-cell

The Donnan membrane equilibrium

Certain complications arise when solutions containing both non-diffusible and (inevitably) diffusible ionic species are considered. Gibbs predicted and later Donnan demonstrated that when the non-diffusible ions are located on one side of a semipermeable membrane, the distribution of the diffusible ions is unequal when equilibrium is attained, being greater on the side of the membrane containing the non-diffusible ions. This distribution can be calculated thermodynamically, although a simpler kinetic treatment will suffice.

Consider a simple example in which equal volumes of solutions of the sodium salt of a protein and of sodium chloride with respective equivalent concentrations a and b are initially separated by a semipermeable membrane, as shown in Figure 2.11. To maintain overall electrical neutrality Na^+ and Cl^- ions must diffuse across the membrane in pairs. The rate of diffusion in any particular direction will depend on the probability of an Na^+ and a Cl^- ion arriving at a given point on the membrane surface simultaneously. This probability is

	(1)	(2)
Initial concentrations	$Na^+ = a$ $Pr^- = a$	$Na^+ = b$ $Cl^- = b$
Equilibrium concentrations	$Na^+ = a + x$ $Pr^- = a$ $Cl^- = x$	$Na^+ = b - x$ $Cl^- = b - x$

Figure 2.11 The Donnan membrane equilibrium

proportional to the product of the Na^+ and Cl^- ion concentrations (strictly, activities), so that

$$\text{rate of diffusion from (1) to (2)} = k(a+x)x$$

$$\text{rate of diffusion from (2) to (1)} = k(b-x)^2$$

At equilibrium these rates of diffusion are equal – i.e.

or $(a+x)x = (b-x)^2$

$$x = \frac{b^2}{a+2b}$$

At equilibrium the concentrations of diffusible ions in compartments (1) and (2) are $(a+2x)$ and $2(b-x)$, respectively, so that the excess concentration in compartment (1) is $(a-2b+4x)$. Substituting for x, this excess diffusible ion concentration works out to be $a^2/(a+2b)$.

Clearly, the results of osmotic pressure measurements on solutions of charged colloidal particles, such as proteins, will be invalid unless precautions are taken either to eliminate or to correct for this Donnan effect. Working at the isoelectric pH of the protein will eliminate the Donnan effect but will probably introduce new errors due to coagulation of the protein. Working with a moderately large salt concentration and a small protein concentration will make the ratio $a^2/(a+2b)$ small and allow the Donnan effect to be virtually eliminated.

Rotary Brownian motion

In addition to translational Brownian motion, suspended molecules or particles undergo random rotational motion about their axes, so that, in the absence of aligning forces, they are in a state of random orientation. Rotary diffusion coefficients can be defined (ellipsoids of revolution have two such coefficients representing rotation about each principal axis) which depend on the size and shape of the molecules or particles in question[28].

Under the influence of an orientating force partial alignment of asymmetric particles takes place, which represents a balance between the aligning force on the particles and their rotary diffusion. The system will therefore become anisotropic. It is possible to draw conclusions concerning particle dimensions by studying various changes in physical properties brought about by such particle alignment.

Streaming birefringence[35]

The sample is subjected to a strong velocity gradient – for example, in a concentric cylinder viscometer (Figure 2.12) – and the resulting molecular or particle alignment causes the previously isotropic solution to become doubly refracting (birefringent). The magnitude of the birefringence is related in an elaborate theory to the rotational diffusion coefficient, and, hence, to the molecular or particle dimensions.

In certain cases (e.g. iron(III) hydroxide sol) birefringence can be produced by the aligning action of electrical or magnetic fields.

Dielectric dispersion

When a solution containing dipolar molecules is placed between electrodes and subjected to an alternating current, the molecules tend to rotate in phase with the current, thus increasing the dielectric constant of the solution. As the frequency is increased, it becomes more difficult for the dipolar molecules to overcome the viscous

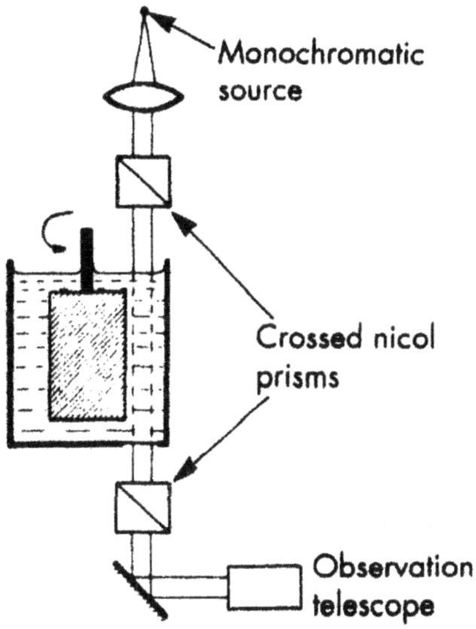

Figure 2.12 Apparatus for observing streaming birefringence

resistance of the medium rapidly enough to remain in phase, and the dielectric constant drops in a more or less stepwise fashion. Each characteristic frequency where there is a notable change of dielectric constant is related to the time taken for the molecule to rotate about a particular axis, and hence, to the appropriate rotary diffusion coefficient.

3 Optical properties

Optical and electron microscopy

The optical microscope – resolving power

Colloidal particles are often too small to permit direct microscopic observation. The resolving power of an optical microscope (i.e. the smallest distance by which two objects may be separated and yet remain distinguishable from each other) is limited mainly by the wavelength λ of the light used for illumination. The limit of resolution δ is given by the expression

$$\delta = \lambda / 2n \sin \alpha \tag{3.1}$$

where α is the angular aperture (half the angle subtended at the object by the objective lens), n is the refractive index of the medium between the object and the objective lens, and $n \sin \alpha$ is the numerical aperture of the objective lens for a given immersion medium.

The numerical aperture of an optical microscope is generally less than unity. With oil-immersion objectives numerical apertures up to about 1.5 are attainable, so that, for light of wavelength 600 nm, this would permit a resolution limit of about 200 nm (0.2 μm). Since the human eye can readily distinguish objects some 0.2 mm (200 μm) apart, there is little advantage in using an optical microscope, however well constructed, which magnifies more than about 1000 times. Further magnification increases the size but not the definition of the image.

Owing to its large numerical aperture, the depth of focus of an optical microscope is relatively small (*c.* 10 μm at × 100 magnification and *c.* 1 μm at × 1000 magnification). This is not always a disadvantage of the technique; for example, in microelectrophoresis (see page 192) it permits the observation of particles located at a narrowly defined level in the electrophoresis cell.

Particle sizes as measured by optical microscopy are likely to be in serious error for diameters less than *c.* 2 μm, although the limit of resolution is some ten times better than this (see Table 3.1).

Table 3.1 Determination of the diameters of spherical particles by optical microscopy[29]

True diameter/μm	*Visual estimate*/μm
1.0	1.13
0.5	0.68
$\leqslant$ 0.2	0.5

In addition to the question of resolving power, the visibility of an object may be limited owing to lack of optical contrast between the object and its surrounding background.

Two techniques for overcoming the limitations of optical microscopy are of particular value in the study of colloidal systems. They are electron microscopy[36–37], in which the limit of resolution is greatly extended, and dark-field microscopy, in which the minimum observable contrast is greatly reduced.

The transmission electron microscope

To increase the resolving power of a microscope so that matter of colloidal (and smaller) dimensions may be observed directly, the wavelength of the radiation used must be reduced considerably below that of visible light. Electron beams can be produced with wavelengths of the order of 0.01 nm and focused by electric or magnetic fields, which act as the equivalent of lenses. The resolution of an electron microscope is limited not so much by wavelength as by the technical difficulties of stabilising high-tension supplies and correcting lens aberrations. Only lenses with a numerical aperture of less than 0.01 are usable at present. With computer application to smooth out 'noise' a resolution of 0.2 nm has been attained, which compares with atomic dimensions. Single atoms, however, will appear blurred

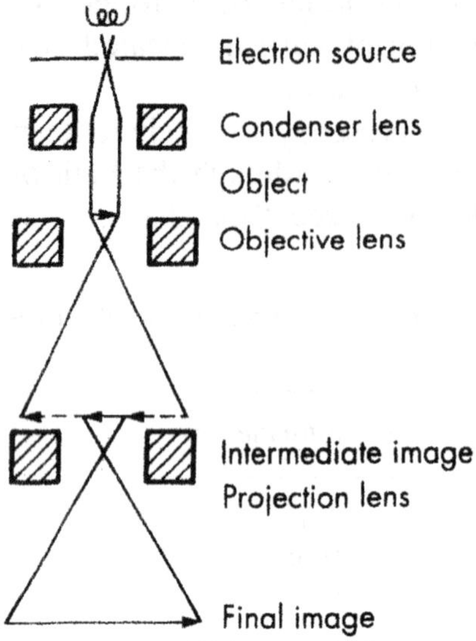

Figure 3.1 Schematic representation of the transmission electron microscope

irrespective of the resolution, owing to the rapid fluctuation of their location.

The useful range of the transmission electron microscope for particle size measurement is *c.* 1 nm–5 μm diameter. Owing to the complexity of calculating the degree of magnification directly, this is usually determined by calibration using characterised polystyrene latex particles or a diffraction grating.

The use of the electron microscope for studying colloidal systems is limited by the fact that electrons can only travel unhindered in high vacuum, so that any system having a significant vapour pressure must be thoroughly dried before it can be observed. Such pretreatment may result in a misrepresentation of the sample under consideration. Instability of the sample to electron beams could also result in misrepresentation.

A small amount of the material under investigation is deposited on an electron-transparent plastic or carbon film (10–20 nm thick) supported on a fine copper mesh grid. The sample scatters electrons out of the field of view, and the final image can be made visible on a fluorescent screen. The amount of scattering depends on the thickness and on the atomic number of the atoms forming the specimen, so that organic materials are relatively electron-transparent and show little contrast against the background support, whereas materials containing heavy metal atoms make ideal specimens.

To enhance contrast and obtain three-dimensional effects, the technique of shadow-casting is generally employed. A heavy metal, such as gold, is evaporated in vacuum and at a known angle on to the specimen, which gives a side illumination effect (see Figure 3.2). From the angle of shadowing and the length of the shadows, a three-dimensional picture of the specimen can be built up. An even better picture can be obtained by lightly shadowing the sample in two directions at right angles.

A most useful technique for examining surface structure is that of replication. One method is to deposit the sample on a freshly cleaved mica surface on to which carbon (and, if desired, a heavy metal) is vacuum-evaporated. The resulting thin film, with the specimen particles still embedded, is floated off the mica on to a water surface. The particles are dissolved out with a suitable solvent and the resulting replica is mounted on a copper grid.

The scanning electron microscope

In the scanning electron microscope a fine beam of medium-energy electrons scans across the sample in a series of parallel tracks. These interact with the sample to produce various signals, including secondary electron emission (SEE), back-scattered electrons (BSE), cathodoluminescence and X-rays, each of which (with their varying characteristics) can be detected, displayed on a fluorescent screen and photographed. In the SEE mode the particles appear to be diffusely illuminated, particle size can be measured and aggregation behaviour can be studied, but there is little indication of height. In the BSE mode the particles appear to be illuminated from a point source and the resulting shadows lead to a good impression of height.

The magnification achieved in a scanning electron microscope (resolution limit of *c.* 5 nm) is, in general, less than that in a transmission electron microscope, but the major advantage of the technique (which is a consequence of the low numerical aperture) is the great depth of focus which can be achieved. At magnifications in the range of optical microscopy the scanning electron microscope can give a depth of focus several hundred times greater than that of the optical microscope. In colloid and surface science this large depth of

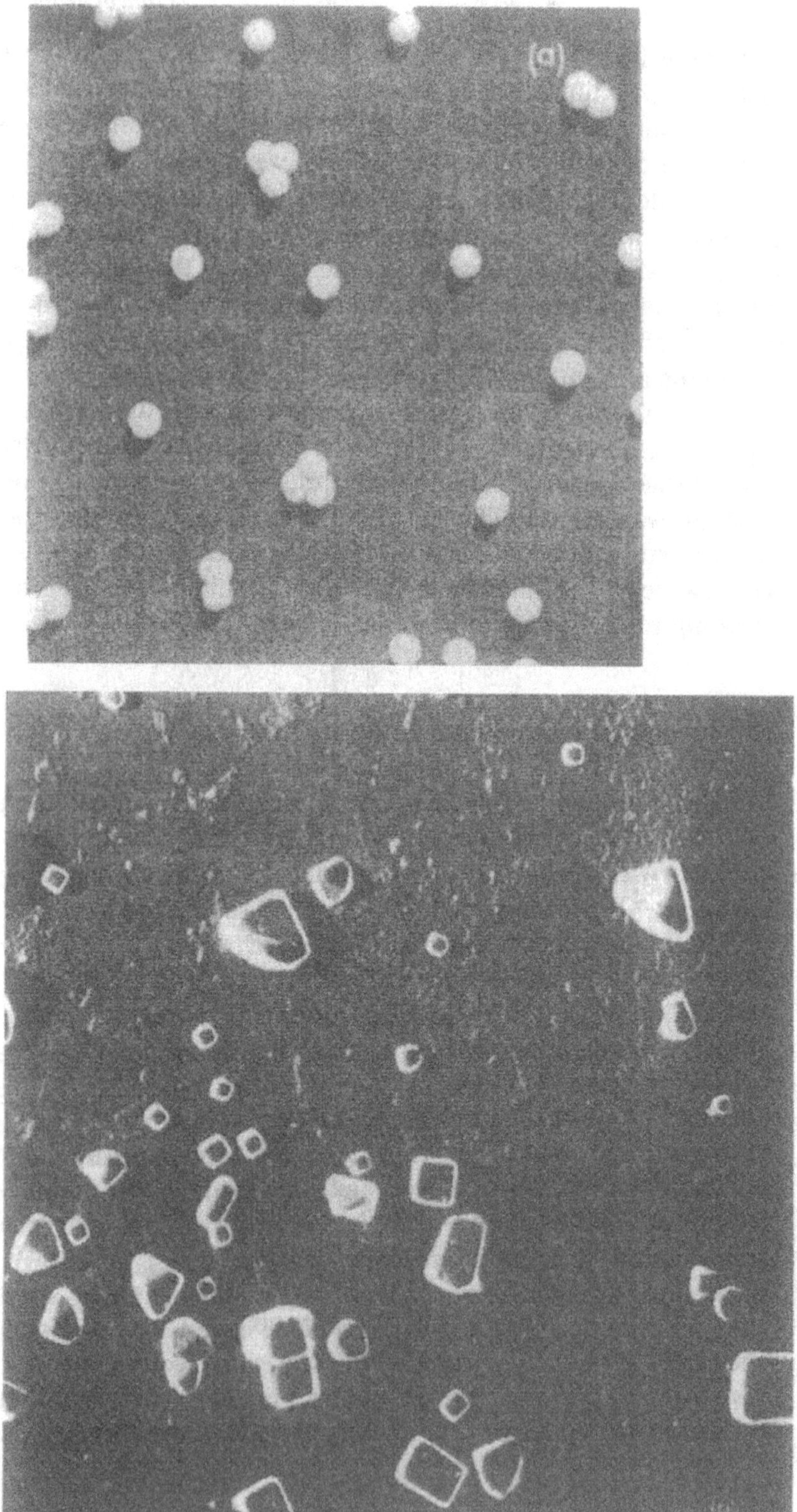

Figure 3.2 Electron micrographs. (a) Shadowed polystyrene latex particles (×50 000). (b) Shadowed silver chloride particles (× 15 000)

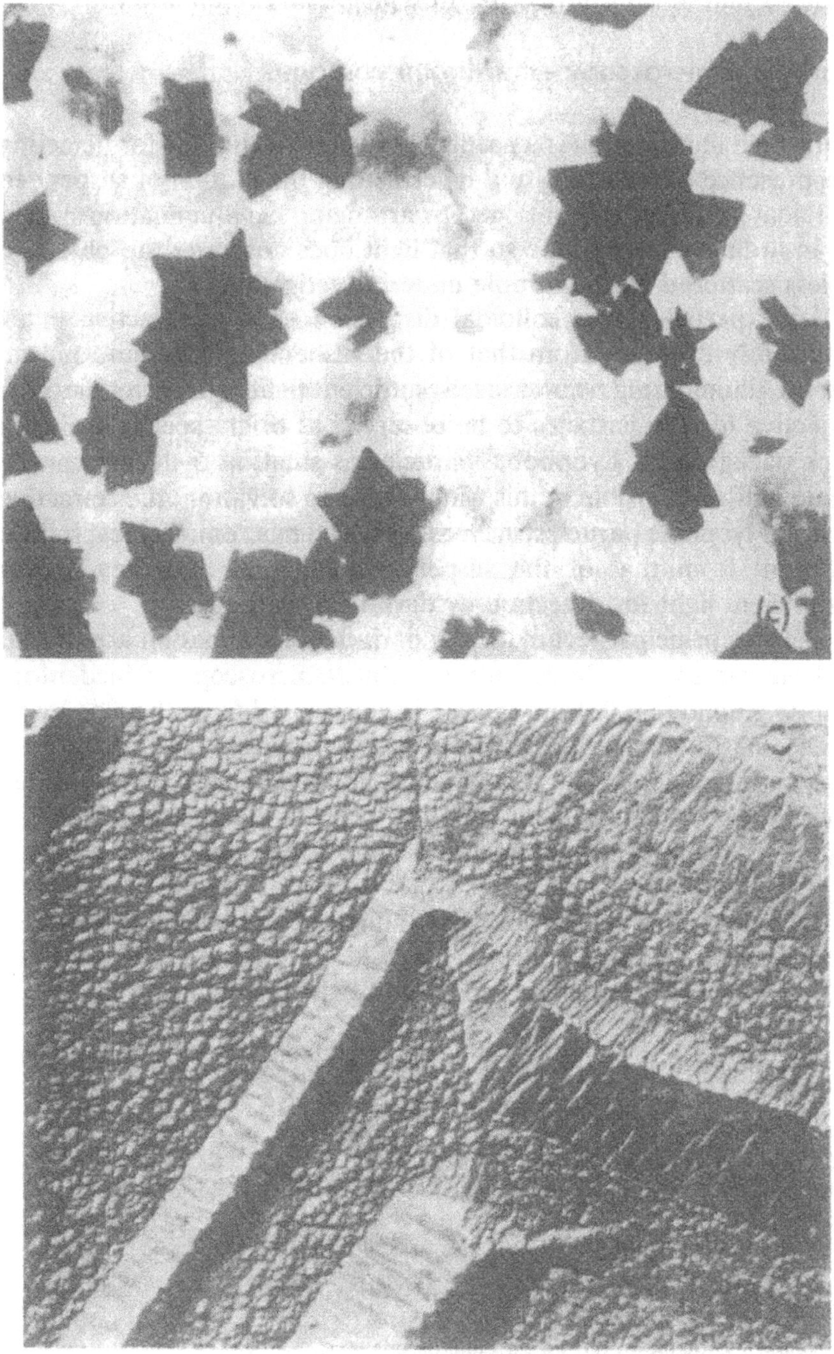

Figure 3.2 Electron micrographs. (c) Platelets of nordstrandite (aluminium hydroxide) (× 5000). (d) Replica of an etched copper surface (× 5600)

focus is extremely valuable in the study of the contours of solid surfaces and in the study of particle shape and orientation.

Dark-field microscopy – the ultramicroscope

Dark-field illumination is a particularly useful technique for detecting the presence of, counting and investigating the motion of suspended colloidal particles. It is obtained by arranging the illumination system of an ordinary microscope so that light does not enter the objective unless scattered by the sample under investigation.

If the particles in a colloidal dispersion have a refractive index sufficiently different from that of the suspending medium, and an intense illuminating beam is used, sufficient light is deflected into the objective for the particles to be observed as bright specks against a dark background. Lyophobic particles as small as 5–10 nm can be made indirectly visible in this way. Owing to solvation, the refractive index of lyophilic particles, such as dissolved macromolecules, is little different from that of the suspending medium, and they scatter insufficient light for detection by dark-field methods.

The two principal techniques of dark-field illumination are the slit and the cardioid methods. In the slit ultramicroscope of Siedentopf and Zsigmondy (1903) the sample is illuminated from the side by an intense narrow beam of light from a carbon-arc source (Figure 3.3). The cardioid condenser (a standard microscope accessory) is an optical device for producing a hollow cone of illuminating light; the sample is located at the apex of the cone, where the light intensity is high (Figure 3.4).

Dark-field methods do not help to improve the resolving power of a microscope. A small scattering particle is seen indirectly as a weak blur. Two particles must be separated by the resolution distance δ to

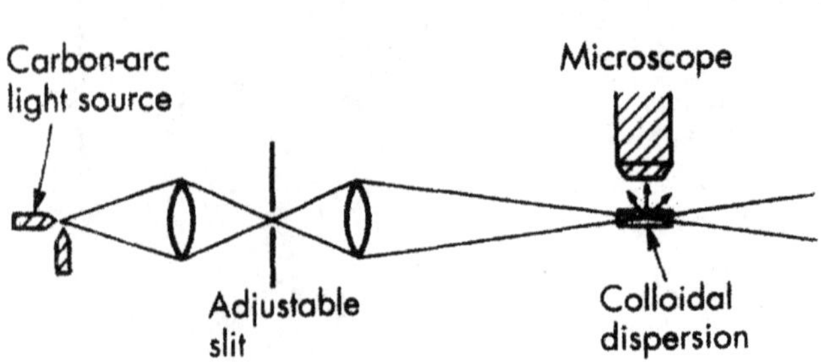

Figure 3.3 Principle of the slit ultramicroscope

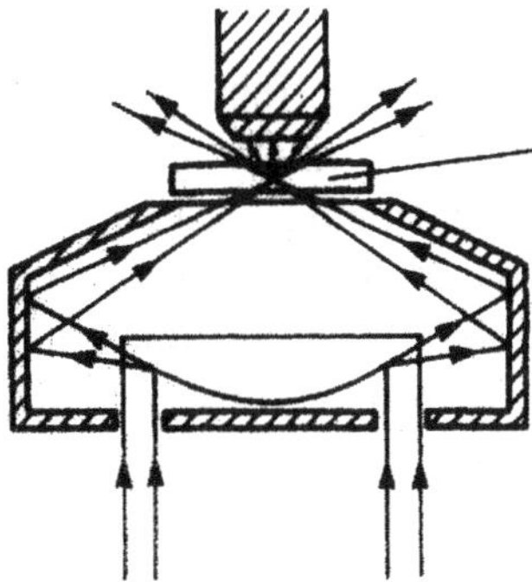

Figure 3.4 Principle of the cardioid dark-field condenser

be separately visible. Dark-field microscopy is, nevertheless, an extremely useful technique for studying colloidal dispersions and obtaining information concerning:

1. Brownian motion.
2. Sedimentation equilibrium.
3. Electrophoretic mobility.
4. The progress of particle aggregation.
5. Number-average particle size (from counting experiments and a knowledge of the concentration of dispersed phase).
6. Polydispersity (the larger particles scatter more light and therefore appear to be brighter).
7. Asymmetry (asymmetric particles give a flashing effect, owing to different scattering intensities for different orientations).

Light scattering[38–40]

When a beam of light is directed at a colloidal solution or dispersion, some of the light may be absorbed (colour is produced when light of certain wavelengths is selectively absorbed), some is scattered and the remainder is transmitted undisturbed through the sample. Light scattering results from the electric field associated with the incident light inducing periodic oscillations of the electron clouds of the atoms of the material in question – these then act as secondary sources and radiate scattered light.

The Tyndall effect – turbidity

All materials are capable of scattering light (Tyndall effect) to some extent. The noticeable turbidity associated with many colloidal dispersions is a consequence of intense light scattering. A beam of sunlight is often visible from the side because of light scattered by dust particles. Solutions of certain macromolecular materials may appear to be clear, but in fact they are slightly turbid because of weak light scattering. Only a perfectly homogeneous system would not scatter light; therefore, even pure liquids and dust-free gases are very slightly turbid.

The turbidity of a material is defined by the expression

$$I_t / I_0 = \exp[-\tau l] \tag{3.2}$$

where I_0 is the intensity of the incident light beam, I_t is the intensity of the transmitted light beam, l is the length of the sample and τ is the turbidity.

Measurement of scattered light

As we shall see, the intensity, polarisation and angular distribution of the light scattered from a colloidal system depend on the size and shape of the scattering particles, the interactions between them, and the difference between the refractive indices of the particles and the dispersion medium. Light-scattering measurements are, therefore, of great value for estimating particle size, shape and interactions, and have found wide application in the study of colloidal dispersions, association colloids, and solutions of natural and synthetic macromolecules.

Light scattering offers the following advantages over some of the alternative techniques of particle-size analysis:

1. It is absolute – no calibration is required.
2. Measurements are made almost instantaneously, which makes it suitable for rate studies.
3. There is no significant perturbation of the system.
4. The number of particles involved is very large, which permits representative sampling of polydispersed samples.

The intensity of the light scattered by colloidal solutions or dispersions of low turbidity is measured directly. A detecting photocell is mounted on a rotating arm to permit measurement of the light scattered at several angles, and fitted with a polaroid for observing the polarisation of the scattered light (see Figure 3.5). Weakening of the scattered beam itself as it passes through the slightly turbid sample can be neglected, and its intensity can be compared with that of the transmitted beam.

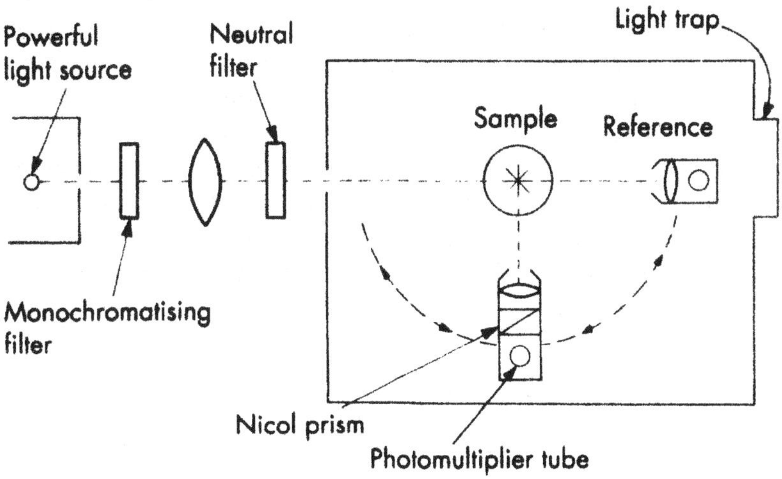

Figure 3.5 Measurement of scattered light

Although simple in principle, light-scattering measurements present a number of experimental difficulties, the most notable being the necessity to free the sample from impurities such as dust, the relatively large particles of which would scatter light strongly and introduce serious errors.

Light-scattering theory

It is convenient to divide the scattering of light by independent particles into three classes:

1. Rayleigh scattering (where the scattering particles are small enough to act as point sources of scattered light).

2. Debye scattering (where the particles are relatively large, but the difference between their refractive index and that of the dispersion medium is small).
3. Mie scattering (where the particles are relatively large and have a refractive index significantly different from that of the dispersion medium).

Scattering by small particles

Rayleigh (1871) laid the foundation of light-scattering theory by applying the electromagnetic theory of light to the scattering by small, non-absorbing (insulating), spherical particles in a gaseous medium. When an electromagnetic wave of intensity I_0 and wavelength λ falls on a small ($< c.\ \lambda/20$) particle of polarisability α, oscillating dipoles are induced in the particle. The particle then serves as a secondary source for the emission of scattered radiation of the same wavelength as the incident light. For an unpolarised incident beam, the intensity I_θ at a distance r from the particle of the light scattered at an angle θ to the incident beam is given by the expression

$$\frac{I_\theta r^2}{I_0} = \frac{8\pi^4\alpha^2}{\lambda^4}(1+\cos^2\theta) = R_\theta(1+\cos^2\theta) \tag{3.3}$$

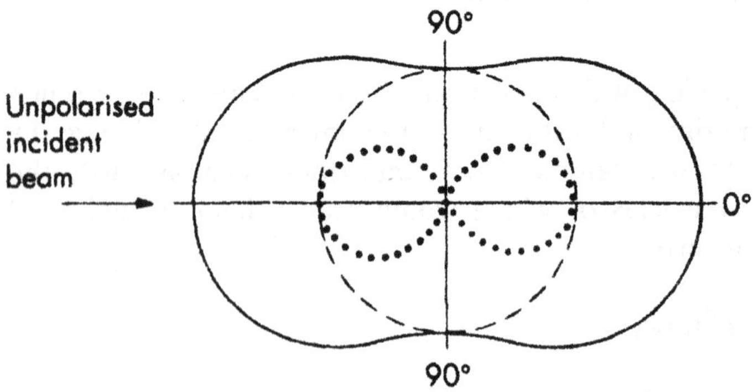

Figure 3.6 Radiation envelope for light scattered from small particles. Distances from the origin of the dotted, dashed and smooth lines represent the relative intensities of the horizontally polarised component, vertically polarised component and total scattered light, respectively

The quantity R_θ $(1+\cos^2\theta)$ is called the Rayleigh ratio. The unity term in $(1+\cos^2\theta)$ refers to the vertically polarised component of the scattered light, and the $\cos^2\theta$ term to the horizontally polarised component.

Since the scattering intensity is proportional to $1/\lambda^4$, blue light ($\lambda \sim$ 450 nm) is scattered much more than red light ($\lambda \sim$ 650 nm). With incident white light, a scattering material will, therefore, tend to appear to be blue when viewed at right angles to the incident beam and red when viewed from end-on. This phenomenon is evident in the blue colour of the sky, tobacco smoke, diluted milk, etc., and in the yellowish-red of the rising and setting sun.

Interparticle interference

If the scattering sources in a system are close together and regularly spaced, as in a crystalline material, there will be regular phase relationships (coherent scattering) and, therefore, almost total destructive interference between the scattered light waves – i.e. the intensity of the resulting scattered light will be almost zero. When the scattering sources are randomly arranged, which is virtually the case for gases, pure liquids and dilute solutions or dispersions, there are no definite phase relationships (incoherent scattering) and destructive interference between the scattered light waves is incomplete.

For a system of independent scatterers (point sources of scattered light distributed completely at random), the emitted light waves have an equal probability of reinforcing or destructively interfering with one another. The amplitudes of the scattered waves add and subtract in a random fashion, with the result that (by analogy with Brownian displacement; page 25) the amplitude of the total scattered light is proportional to the square root of the number of scattering particles. Since the intensity of a light wave is proportional to the square of its amplitude, the total intensity of scattered light is proportional to the number of particles.

Relative molecular masses from light-scattering measurements

If the dimensions of a scattering particle are all less than *c*. $\lambda/20$, the scattered light waves emanating from the various parts of the particle cannot be more than *c*. $\lambda/10$ out of phase, and so their amplitudes are practically additive. The total amplitude of the light scattered from such a particle is, therefore, proportional to the number of individual scatterers in the particle – i.e. to its volume and, hence, its mass; and the total intensity of scattered light is proportional to the square of the particle mass. Consequently, for a random dispersion containing n particles of mass m, the total amount of light scattered is proportional to nm^2; and as nm is proportional to the concentration c of the dispersed phase,

$$\text{total light scattered} \propto cm$$

An alternative (but equivalent) approach is the so-called fluctuation theory, in which light scattering is treated as a consequence of random non-uniformities of concentration and, hence, refractive index, arising from random molecular movement (see page 26). Using this approach, the above relationship can be written in the quantitative form derived by Debye[140] for dilute macromolecular solutions:

$$\frac{Hc}{\tau} = \frac{1}{M} + 2Bc$$

$$\text{i.e.} \quad \left(\frac{Hc}{\tau}\right)_{\lim c \to 0} = \frac{1}{M} \tag{3.4}$$

where τ is the turbidity of the solution, M is the molar mass of the solute, B is the same as B_2 in equation (2.21) and H is a constant given by

$$H = \frac{32\pi^3 n_0^2}{3N_A \lambda_0^4}\left(\frac{dn}{dc}\right)^2 \tag{3.5}$$

where n_0 is the refractive index of the solvent, n is the refractive index of the solution and λ_0 is the wavelength *in vacuo* (i.e. $\lambda_0 = n\lambda$, where λ is the wavelength of the light in the solution). τ is calculated from the intensity of light scattered at a known angle (usually 90° or 0°). Summation of the products, R dω, over the solid angle 4π leads to the relationship

$$\tau = \frac{16\pi}{3} R_{90^\circ} \tag{3.6}$$

where R (defined in equation 3.3) now refers to primary scattering from unit volume of solution.

Therefore,

$$\left(\frac{Kc}{R_{90^\circ}}\right)_{\lim c\to 0} = \frac{1}{M} \tag{3.7}$$

where

$$K = \frac{2\pi^2 n_0^2}{N_A \lambda_0^4}\left(\frac{\mathrm{d}n}{\mathrm{d}c}\right)^2 \tag{3.8}$$

$\mathrm{d}n/\mathrm{d}c$ is measured with a differential refractometer reading to the sixth decimal place.

In contrast to osmotic pressure, light-scattering measurements become easier as the particle size increases. For spherical particles the upper limit of applicability of the Debye equation is a particle diameter of *c.* $\lambda/20$ (i.e. 20–25 nm for $\lambda_0 \sim 600$ nm or $\lambda_{\text{water}} \sim 450$ nm; or a relative molecular mass of the order of 10^7). For asymmetric particles this upper limit is lower. However, by modification of the theory, much larger particles can also be studied by light scattering methods. For polydispersed systems a mass-average relative molecular mass is given.

Large particles

The theory of light scattering is more complicated when one or more of the particle dimensions exceeds *c.* $\lambda/20$. Such particles cannot be considered as point sources of scattered light, and destructive interference between scattered light waves originating from different locations on the same particle must be taken into account. This intraparticle destructive interference is zero for light scattered in a forward direction ($\theta = 0°$). Extinction will take place between waves scattered backwards from the front and rear of a spherical particle of diameter $\lambda/4$ (i.e. the total path difference is $\lambda/2$). The radiation envelope for such a particle will, therefore, be unsymmetrical, more light being scattered forwards than backwards.

When particles of refractive index significantly different from that of the suspending medium contain a dimension greater than *c.* $\lambda/4$, extinction at intermediate angles is possible and maxima and minima of scattering can be observed at different angles. The location of such maxima and minima will depend on the wavelength, so that with

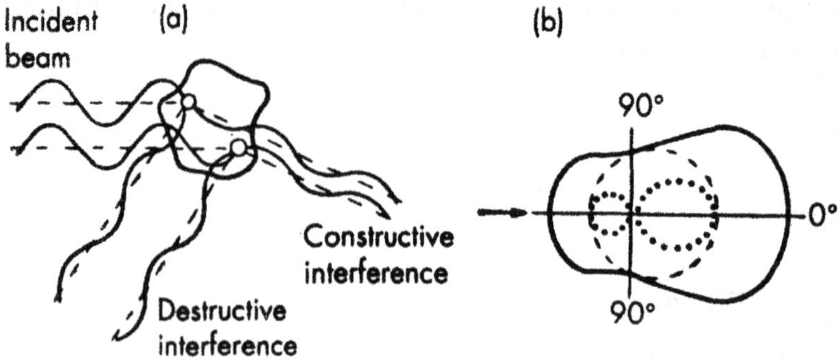

Figure 3.7 (a) Scattering from a relatively large particle. (b) Radiation envelope for light scattered from a spherical particle ($x = 0.8$, $m = 1.25$). *See* text and *Figure 3.6* for explanation

incident white light it is possible, with a suitable monodispersed system, to observe spectral colour sequences (known as 'higher-order Tyndall spectra').

Mie (1908) elaborated a general quantitative theory for light scattering by spherical particles. The intensity of the light scattered at various angles is related to m, the ratio of the refractive index of the particles to that of the dispersion medium, and the parameter $x = 2\pi r/\lambda$. Mie's theory has been extended by Gans to include certain non-spherical shapes. The main features of Mie's theory were verified by La Mer and Barnes[141] from measurements of the angular variation of the light scattered by monodispersed sulphur sols containing particles of radius 300–600 nm.

Since the light scattered forwards (0°) suffers no intraparticle interference, its intensity is proportional to the square of the particle mass. By measuring the light scattered by a colloidal solution or dispersion as a function of both angle and concentration and extrapolating to zero angle and zero concentration, the size of relatively large particles can be calculated from the Debye equation. This extrapolation (Zimm plot[142]; Figure 3.8) is achieved by plotting Kc/R_θ against $\sin^2(\theta/2) + kc$, where k is an arbitrary constant selected so as to give convenient spacing between the points on the graph.

$$\left.\frac{Kc}{R_\theta}\right|_{\lim c\to 0,\ \theta\to 0} = \frac{1}{M} \tag{3.9}$$

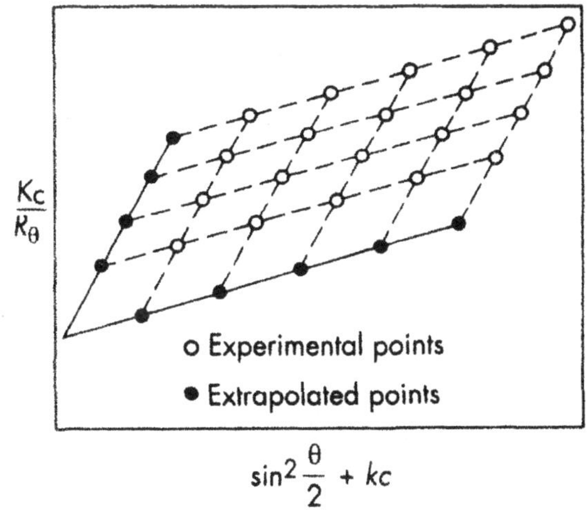

Figure 3.8 A Zimm plot

Particles which are too small to show a series of maxima and minima in the angular variation of scattered light are frequently studied by measuring the dissymmetry of scattering (usually defined as the ratio of the light scattered at 45° to that scattered at 135°). The dissymmetry of scattering is a measure of the extent of the particles compared with λ. If the molecular or particle size is known, it can be related to the axial ratio of rod-like particles or the coiling of flexible linear macromolecules.

The application of light scattering to the characterisation of colloidal systems has advanced rapidly over the last few decades. This has been made possible by the development of (*a*) lasers as intense, coherent and well-collimated light sources, (*b*) sophisticated electronic devices for recording data, and (*c*) computers for the complex data processing that is involved.

Dynamic light scattering

The precisely defined frequencies associated with laser sources makes it possible to exploit light scattering to study the motion of colloidal particles. Light scattered by a moving particle will experience a Doppler shift to slightly higher or lower frequency depending on whether the particle is moving towards or away from the observer. For a collection of particles moving at random by virtue of their Brownian motion, a Doppler frequency broadening will result.

Mixing this broadened, scattered signal with incident light produces a pattern of beat frequencies, the measurement of which allows the diffusion coefficient of the particles to be calculated.

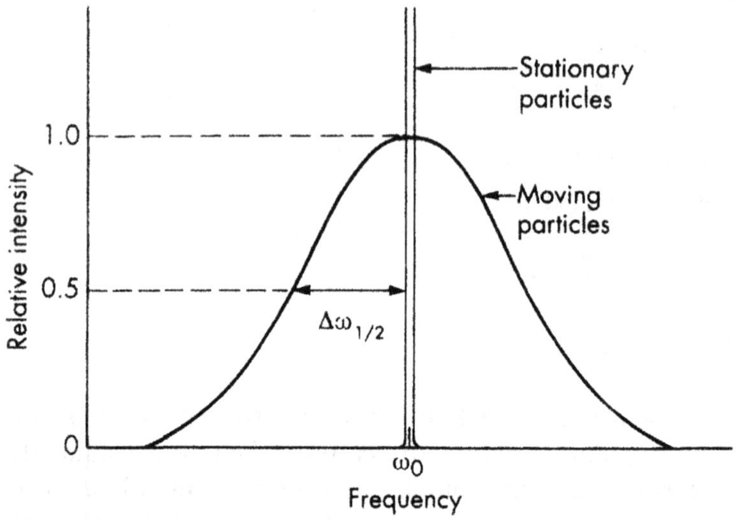

Figure 3.9 Doppler broadening. For a typical colloidal dispersion, $\Delta\omega_{1/2} \approx 10^3$ with $\omega_0 \approx 10^{14}$ Hz

The Doppler shift will vary from zero for zero scattering angle to maximum for back-scattering (θ = 180°). The width of the Doppler-broadened peak at half-height (Figure 3.9) is related to the diffusion coefficient by

$$\Delta\omega_{1/2} = D\left(\frac{4\pi n_0 \omega_0}{c}\sin\frac{\theta}{2}\right)^2 \tag{3.10}$$

where θ is the scattering angle.

The technique is alternatively called photon correlation spectroscopy (PCS) or quasi-elastic light scattering (QELS).

An interesting modification of this technique is the fibre-optic dynamic anemometer (FODA)[143]. A length of fibre-optic cable carries the laser beam to the interior of the dispersion. Back-scattered light, with its Doppler frequency shift, is returned to the detector along with reflected light and, again, the resulting beat frequency pattern is analysed. Since only a very small volume around the fibre-optic tip is sampled, the technique can be used for concentrated dispersions.

Small angle neutron scattering

Neutrons from reactors or accelerators are slowed down in a moderator to kinetic energies corresponding to room temperature (thermal neutrons) or less (cold neutrons), for which the de Broglie wavelengths ($\lambda = h/mv$) are of the order of 10^{-10} to 10^{-9} m. Therefore, $a/\lambda \gg 1$. The theory of light scattering by large particles, discussed on pages 59–61, applies equally to neutron scattering, except that the dependence on the relative values of the refractive indices of particles and dispersion medium is replaced by a corresponding dependence on neutron scattering length densities. In contrast to conventional light scattering, neutron scattering is very weak and, consequently, it can be exploited to study concentrated dispersions.

Neutron scattering is also useful for the study of adsorbed material on hydrosol particles. The neutron scattering length densities of hydrogen and deuterium differ considerably. By preparing a hydrosol in an appropriate H_2O/D_2O mixture, it is possible to match the neutron scattering length densities of the dispersion medium and the core particles. The neutron beam thus 'sees' only the adsorbed layer, the thickness of which can be estimated. Alternatively, the dispersion medium can be matched to the adsorbed layer to permit estimation of the core-particle size.

4 *Liquid–gas and liquid–liquid interfaces*

Surface and interfacial tensions

It is well known that short-range forces of attraction exist between molecules (see page 215), and are responsible for the existence of the liquid state. The phenomena of surface and interfacial tension are readily explained in terms of these forces. The molecules which are located within the bulk of a liquid are, on average, subjected to equal forces of attraction in all directions, whereas those located at, for example, a liquid–air interface experience unbalanced attractive forces resulting in a net inward pull (Figure 4.1). As many molecules as possible will leave the liquid surface for the interior of the liquid; the surface will therefore tend to contract spontaneously. For this reason, droplets of liquid and bubbles of gas tend to attain a spherical shape.

Surface tension (and the more fundamental quantity, *surface free energy*) fulfil an outstanding role in the physical chemistry of surfaces.

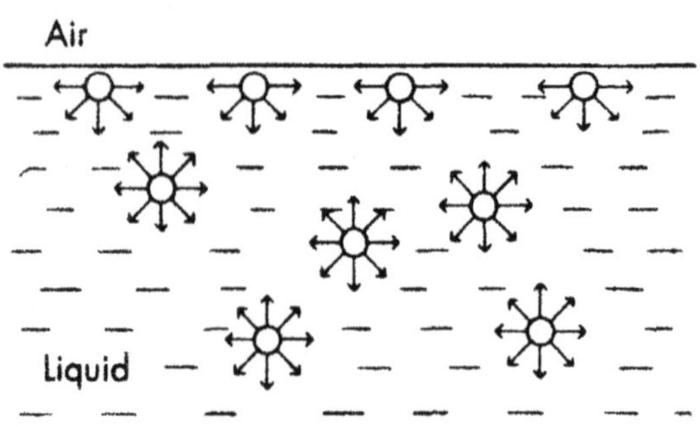

Figure 4.1 Attractive forces between molecules at the surface and in the interior of a liquid

The surface tension γ_0 of a liquid is often defined as the force acting at right angles to any line of unit length on the liquid surface. However, this definition (although appropriate in the case of liquid films, such as in foams) is somewhat misleading, since there is no elastic skin or tangential force as such at the surface of a pure liquid. It is more satisfactory to define surface tension and surface free energy as the work required to increase the area of a surface isothermally and reversibly by unit amount.

There is no fundamental distinction between the terms *surface* and *interface*, although it is customary to describe the boundary between two phases one of which is gaseous as a surface and the boundary between two non-gaseous phases as an interface.

At the interface between two liquids there is again an imbalance of intermolecular forces but of a lesser magnitude. Interfacial tensions usually lie between the individual surface tensions of the two liquids in question.

The above picture implies a static state of affairs. However, it must be appreciated that an apparently quiescent liquid surface is actually in a state of great turbulence on the molecular scale as a result of two-way traffic between the bulk of the liquid and the surface, and between the surface and the vapour phase[41]. The average lifetime of a molecule at the surface of a liquid is *c.* 10^{-6} s.

Table 4.1 Surface tensions and interfacial tensions against water for liquids at 20°C (in mN m^{-1})

Liquid	γ_0	γ_i	*Liquid*	γ_0	γ_i
Water	72.8	–	Ethanol	22.3	–
Benzene	28.9	35.0	*n*-Octanol	27.5	8.5
Acetic acid	27.6	–	*n*-Hexane	18.4	51.1
Acetone	23.7	–	*n*-Octane	21.8	50.8
CCl_4	26.8	45.1	Mercury	485	375

Additivity of intermolecular forces at interfaces

The short-range intermolecular forces which are responsible for surface/interfacial tensions include van der Waals forces (in particular, London dispersion forces, which are universal) and may include hydrogen bonding (as, for example, in water) and metal bonding (as, for example, in mercury). The relatively high values of the surface tensions of water and mercury (see Table 4.1) reflect the contributions of hydrogen bonding and metal bonding, respectively.

These forces are not appreciably influenced by one another, and so may be assumed to be additive. The surface tension of water may, therefore, be considered as the sum of a dispersion force contribution, γ_W^d, and a hydrogen bonding contribution, γ_W^h – i.e.

$$\gamma_W = \gamma_W^d + \gamma_W^h \tag{4.1}$$

Similarly, the surface tension of mercury is made up of dispersion and metal bond contributions:

$$\gamma_{Hg} = \gamma_{Hg}^d + \gamma_{Hg}^m \tag{4.2}$$

In the case of hydrocarbons the surface tension is entirely the result of the dispersion force contribution.

Consider the interface between water and a hydrocarbon oil (see Figure 4.2). Water molecules in the interfacial region are attracted towards the interior of the water phase by water–water interactions (dispersion forces and hydrogen bonding) and towards the oil phase by oil–water interactions (dispersion forces only); likewise, hydrocarbon oil molecules in the interfacial region are attracted to the oil

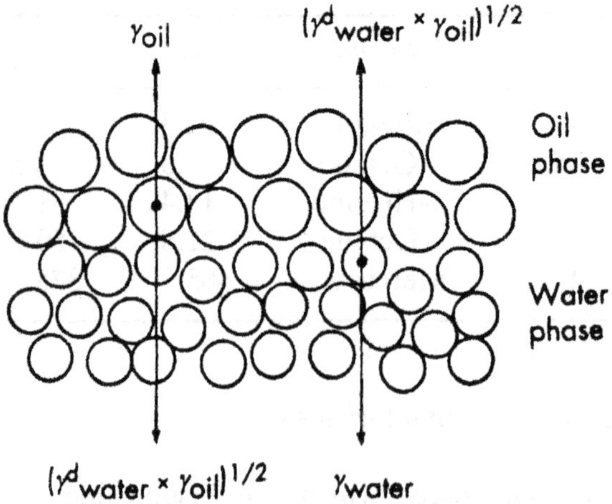

Figure 4.2 Schematic representation of the contributions to an oil–water interfacial tension

phase by oil–oil dispersion forces and to the water phase by oil–water dispersion forces. In a simple approach proposed by Fowkes[42] the oil–water dispersion interactions are considered to be the geometric mean of the oil–oil and water–water dispersion interactions. Hence, the interfacial tension is given by

$$\gamma_{OW} = \gamma_O^d + (\gamma_W^d + \gamma_W^h) - 2 \times (^d{}_W \times \gamma_O^d)^{1/2} \qquad (4.3)$$

Substituting values from Table 4.1 for the *n*-hexane–water interface,

$$51.1 = 18.4 + 72.8 - 2 \times (\gamma_W^d \times 18.4)^{1/2}$$

which gives

$$\gamma_W^d = 21.8 \text{ mNm}^{-1}$$

and

$$\gamma_W^h = 72.8 - 21.8 = 51.0 \text{ mNm}^{-1}$$

Using surface and interfacial tension data for a range of alkanes, Fowkes calculated that $\gamma_W^d = 21.8 \pm 0.7$ mN m^{-1}.

Phenomena at curved interfaces – the Kelvin equation

As a consequence of surface tension, there is a balancing pressure difference across any curved surface, the pressure being greater on the concave side. For a curved surface with principal radii of curvature r_1 and r_2 this pressure difference is given by the Young–Laplace equation, $\Delta p = \gamma(1/r_1 + 1/r_2)$, which reduces to $\Delta p = 2\gamma/r$ for a spherical surface.

The vapour pressure over a small droplet (where there is a high surface/volume ratio) is higher than that over the corresponding flat surface. The transfer of liquid from a plane surface to a droplet requires the expenditure of energy, since the area and, hence, the surface free energy of the droplet will increase.

If the radius of a droplet increases from r to $r + dr$, the surface area will increase from $4\pi r^2$ to $4\pi(r + dr)^2$ (i.e. by $8\pi r\, dr$) and the increase in surface free energy will be $8\pi\gamma r\, dr$. If this process involves the transfer of dn moles of liquid from the plane surface with a vapour pressure p_0 to the droplet with a vapour pressure p_r, the free energy increase is also equal to $\mathrm{d}nRT \ln p_r/p_0$, assuming ideal gaseous behaviour. Equating these free energy increases,

$$\mathrm{d}nRT \ln p_r / p_0 = 8 \pi\gamma r \,\mathrm{d}r$$

and since

$$\mathrm{d}n = 4\pi r^2 \mathrm{d}r \,\rho / M$$

then

$$RT \ln p_r / p_0 = \frac{2\gamma M}{\rho r} = \frac{2\gamma V_m}{r} \qquad (4.4)$$

where ρ is the density of the liquid, V_m is the molar volume of the liquid and M is the molar mass. For example, for water droplets (assuming γ to be constant),

$r = 10^{-7}$ m	$p_r/p_0 \sim 1.01$
10^{-8} m	1.1
10^{-9} m	3.0

This expression, known as the Kelvin equation, has been verified experimentally. It can also be applied to a concave capillary meniscus; in this case the curvature is negative and a vapour pressure lowering is predicted (see page 125).

The effect of curvature on vapour pressure (and, similarly, on solubility) provides a ready explanation for the ability of vapours (and solutions) to supersaturate. If condensation has to take place via droplets containing only a few molecules, the high vapour pressures involved will present an energy barrier to the process, whereas in the presence of foreign matter this barrier can be by-passed.

An important example of this phenomenon is to be found in the ageing of colloidal dispersions (often referred to as Ostwald ripening). In any dispersion there exists a dynamic equilibrium whereby the rates of dissolution and deposition of the dispersed phase balance in order that saturation solubility of the dispersed material in the dispersion medium be maintained. In a polydispersed sol the smaller particles will have a greater solubility than the larger particles and so will tend to dissolve, while the larger particles will tend to grow at their expense. In a sol of a highly insoluble substance, such as silver iodide hydrosol, this phenomenon will be of little consequence, since both large and small particles have extremely little tendency to dissolve. In a sol of more soluble material, such as calcium carbonate hydrosol, however, Ostwald ripening occurs to such an extent that it is not possible to prepare a long-lived dispersion with particles of colloidal dimensions unless a stabilising agent, such as gelatin or a surfactant, is incorporated.

Variation of surface tension with temperature

The surface tension of most liquids decreases with increasing temperature in a nearly linear fashion (some metal melts being exceptional in this respect) and becomes very small in the region of the critical temperature, when the intermolecular cohesive forces approach zero. A number of empirical equations have been suggested which relate surface tension and temperature, one of the most satisfactory being that of Ramsay and Shields:

$$\gamma\left(\frac{Mx}{\rho}\right)^{2/3} = k(T_c - T - 6) \tag{4.5}$$

where M is the molar mass of the liquid, ρ is the density of the liquid, x is the degree of association of the liquid, T_c is the critical temperature and k is a constant.

Measurement of surface and interfacial tensions[2,43,44,144]

The many methods available for the measurement of surface and interfacial tensions can be classified as static, detachment and dynamic, the last of these being used to study relatively short time effects. Static methods usually offer a greater potential for accurate measurement than detachment methods (especially when solutions of surface-active agents are involved)[43], but detachment methods tend to be the more convenient to operate. With careful experimentation and exclusion of contaminants (especially surfactants), it is usually possible to measure surface tensions to an accuracy of 0.01 to 0.1 mN m^{-1}. It is unwise to use water which has been in contact with ion-exchange resins.

Capillary rise method

This is, when properly performed, the most accurate method available for determining surface tensions. Since the measurements do not involve a disturbance of the surface, slow time effects can be followed.

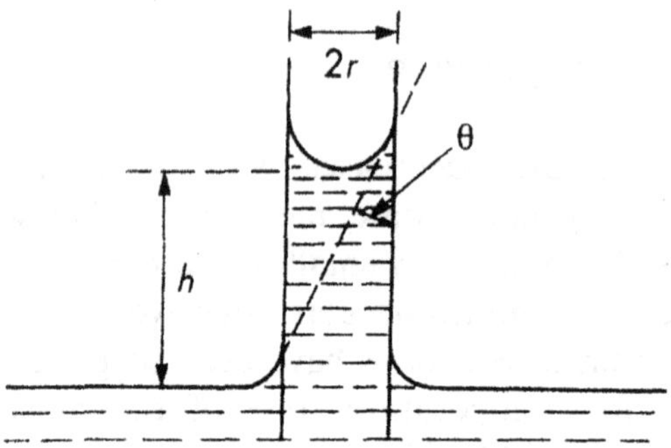

Figure 4.3 Capillary rise

For the rise of a liquid up a narrow capillary

$$\gamma = \frac{rh\Delta\rho g}{2\cos\theta} \tag{4.6}$$

which, for zero contact angle, reduces to

$$\gamma = \tfrac{1}{2}\, rh\Delta\rho g \tag{4.7}$$

where $\Delta\rho$ is (density of liquid − density of vapour).

For accurate work a meniscus correction should be made. In a narrow capillary the meniscus will be approximately hemispherical: therefore,

$$\gamma = \tfrac{1}{2}r(h + r/3)\Delta\rho g \tag{4.8}$$

For wider capillaries one must account for deviation of the meniscus from hemispherical shape[2].

In practice, the capillary rise method is only used when the contact angle is zero, owing to the uncertainty in measuring contact angles correctly. One can check for zero contact angle by allowing the meniscus to approach an equilibrium position in turn from above and below. With a finite contact angle, different equilibrium positions will be noted, owing to differences between advancing and receding contact angles (see page 156). Zero contact angle for aqueous and most other liquids can usually be obtained without difficulty using well-cleaned glass capillaries.

A difficulty associated with this method is that of obtaining capillary tubing of uniform bore (slight ellipticity is not important). Thermometer tubing is useful in this respect. Alternatively, one can adjust the height of the reservoir liquid so as to locate the meniscus at a particular level in the capillary where the cross-sectional area is accurately known.

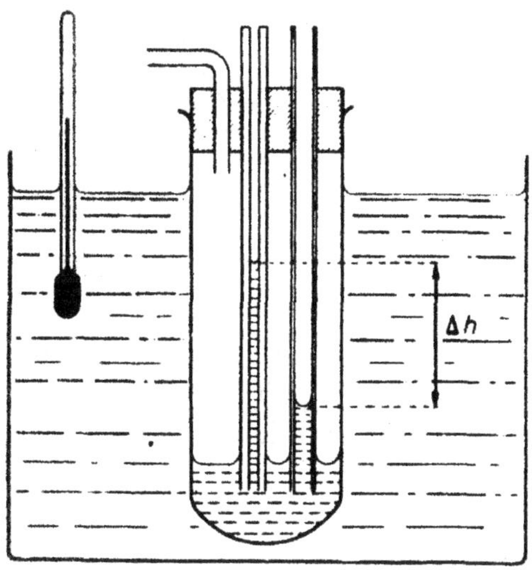

Figure 4.4 Differential capillary rise apparatus

A useful variation is to measure the difference in capillary rise for capillaries of different bore, thus eliminating reference to the flat surface of the reservoir liquid (Figure 4.4). Since $\gamma = \frac{1}{2} r_1 h_1 \Delta\rho g = \frac{1}{2} r_2 h_2 \Delta\rho g$,

$$\gamma = \frac{\Delta\rho g r_1 r_2 \Delta h}{2(r_1 - r_2)} \tag{4.9}$$

Wilhelmy plate methods

A thin mica plate or microscope slide is suspended from the arm of a balance and dips into the liquid, as shown in Figure 4.5.

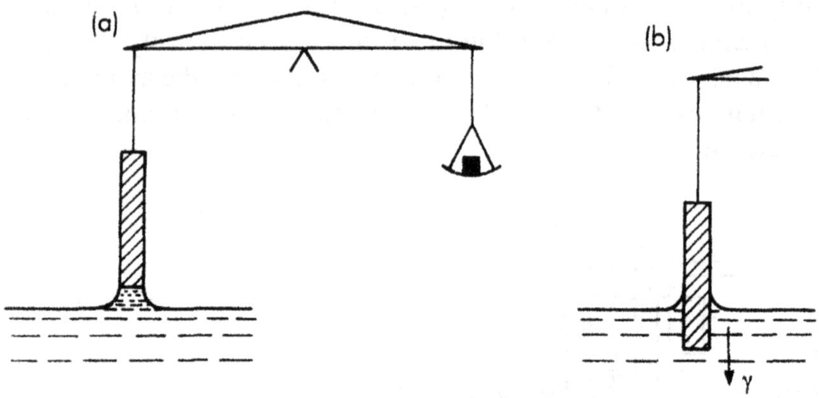

Figure 4.5 Wilhelmy plate methods: (a) detachment; (b) static

When used as a detachment method (Figure 4.5a), the container holding the liquid is gradually lowered and the pull on the balance at the point of detachment is noted. For a plate of length x, breadth y and weight W, assuming zero contact angle,

$$W_{\text{det.}} - W = 2(x + y)\gamma \tag{4.10}$$

The plate method can also be used as a static method (Figure 4.5b) for measuring changes in surface tension (see page 99). The change in the force required to maintain the plate at constant immersion as the surface tension alters is measured.

Ring method

In this method the force required to detach a ring from a surface or interface is measured either by suspending the ring from the arm of a balance or by using a torsion-wire arrangement (du Noüy tensiometer).

The detachment force is related to the surface or interfacial tension by the expression

$$\gamma = \frac{\beta F}{4\pi R} \tag{4.11}$$

where F is the pull on the ring, R is the mean radius of the ring and β is a correction factor.

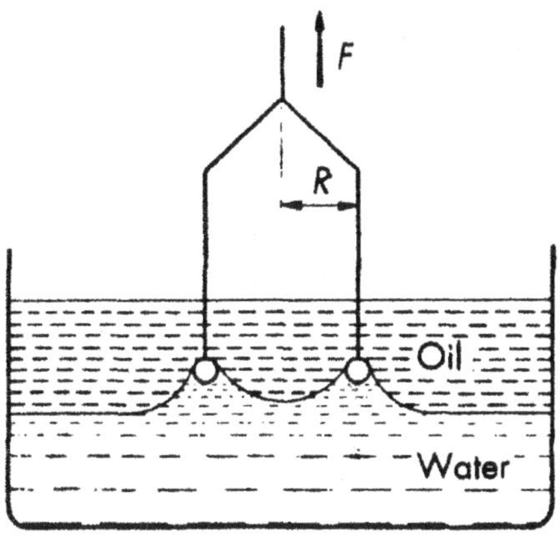

Figure 4.6 Measurement of interfacial tension by the ring method

To ensure zero and, hence, a constant contact angle, platinum rings carefully cleaned in strong acid or by flaming are used. It is essential that the ring should lie flat in a quiescent surface. For interfacial work, the lower liquid must wet the ring preferentially (e.g. for benzene on water, a clean platinum ring is suitable; whereas for water on carbon tetrachloride, the ring must be made hydrophobic).

The correction factor β allows for the non-vertical direction of the tension forces and for the complex shape of the liquid supported by the ring at the point of detachment; hence, it depends on the dimensions of the ring and the nature of the interface. Values of β have been tabulated by Harkins and Jordan[145], they can also be calculated from the equation of Zuidema and Waters[146].

$$(\beta - a)^2 = \frac{4b}{\pi^2} \cdot \frac{1}{R^2} \cdot \frac{F}{4\pi R(\rho_1 - \rho_2)} + c \tag{4.12}$$

where ρ_1 and ρ_2 are the densities of the lower and upper phases; $a = 0.7250$ and $b = 0.090\,75\ \text{m}^{-1}\ \text{s}^2$ for all rings; $c = 0.045\,34–1.679\ r/R$; and r is the radius of the wire.

Drop-volume and drop-weight methods

Drops of a liquid are allowed to detach themselves slowly from the tip of a vertically mounted narrow tube (Figure 4.7) and either they are weighed or their volume is measured. At the point of detachment

$$\gamma = \frac{\phi m g}{2\pi r} = \frac{\phi V \rho g}{2\pi r} \tag{4.13}$$

where m is the mass of the drop, V is the volume of the drop, ρ is the density of the liquid, r is the radius of the tube and ϕ is a correction factor.

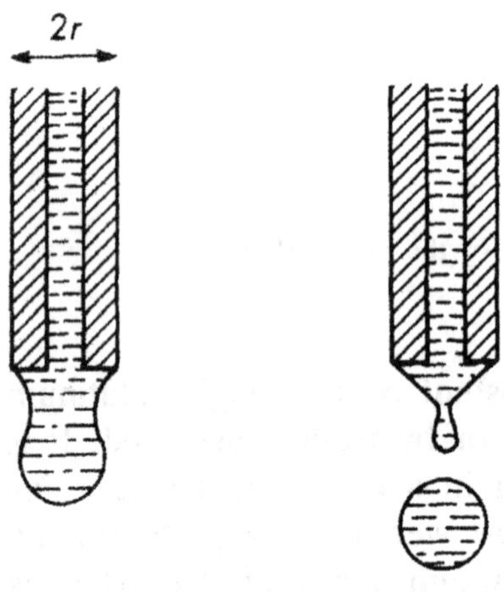

Figure 4.7 Detachment of a drop from the tip of a narrow tube

The correction factor ϕ is required because on detachment (*a*) the drop does not completely leave the tip, (*b*) the surface tension forces are seldom exactly vertical and (*c*) there is a pressure difference across the curved liquid surface[147]. ϕ depends on the ratio $r/V^{1/3}$. Values of ϕ have been determined empirically by Harkins and Brown[148,149]. It can be seen that values of $r/V^{1/3}$ between about 0.6 and 1.2 are preferable (Figure 4.8).

A suitable tip which has been carefully ground smooth used in conjunction with a micrometer syringe burette gives a very convenient

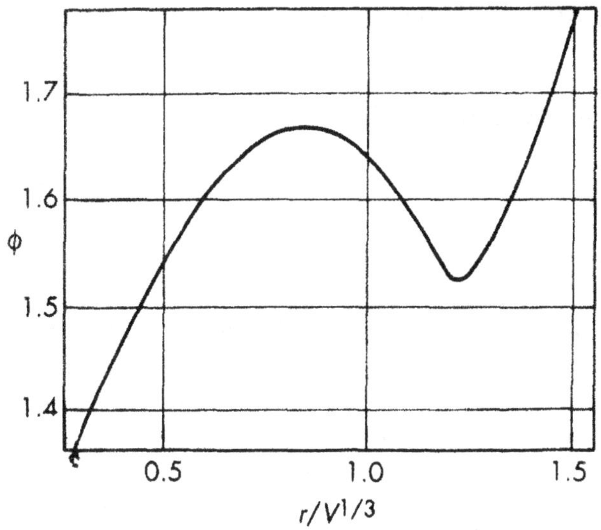

Figure 4.8 Correction factor for drop volume and drop weight methods

drop-volume apparatus for determining both surface and interfacial tensions. The tip of the tube must be completely wetted (r = external radius); alternatively, a tip with a sharp edge can be used. For accurate measurements, the set-up should be free from vibration and the last 10 per cent of the drop should be formed very slowly (~1 min).

Pendant (or sessile) drop profile method

A pendant drop of liquid is photographed or its image is projected on to graph paper. From the various dimensions of the drop the surface or interfacial tension can be computed[2,43,45].

Oscillating jet method[46,150]

This is a dynamic method which enables one to measure the tensions of surfaces at very short times (*c.* 0.01 s) from the moment of their creation. (The methods previously described are used to measure equilibrium tensions.) A jet of liquid emerging from a nozzle of elliptical cross-section is unstable and oscillates about its preferred circular cross-section. Surface tensions can be calculated from the jet dimensions (obtained photographically), flow rate, etc. The age of the surface can be controlled to some extent by altering the flow rate.

Adsorption and orientation at interfaces

Surface activity

Materials such as short-chain fatty acids and alcohols are soluble in both water and oil (e.g. paraffin hydrocarbon) solvents. The hydrocarbon part of the molecule is responsible for its solubility in oil, while the polar –COOH or –OH group has sufficient affinity to water to drag a short-length non-polar hydrocarbon chain into aqueous solution with it. If these molecules become located at an air–water or an oil–water interface, they are able to locate their hydrophilic head groups in the aqueous phase and allow the lipophilic hydrocarbon chains to escape into the vapour or oil phase (Figure 4.9). This situation is energetically more favourable than complete solution in either phase. Further discussion of this point in terms of the intermolecular forces which are involved is given on page 85, where the related phenomenon of micellisation is considered.

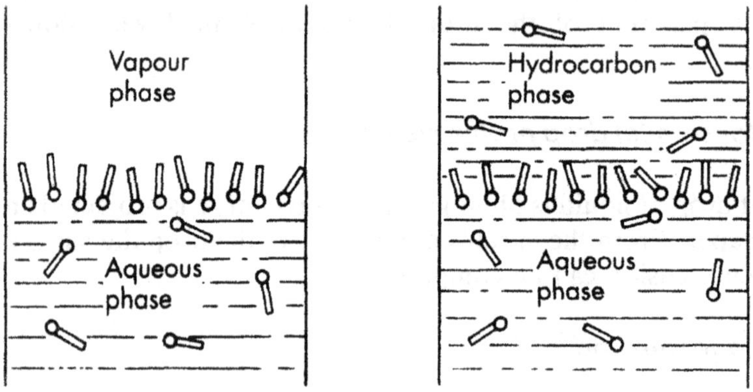

Figure 4.9 Adsorption of surface-active molecules as an orientated monolayer at air–water and oil–water interfaces. The circular part of the molecules represents the hydrophilic polar head group and the rectangular part represents the non-polar hydrocarbon tail. At the air–water interface, the hydrocarbon chains will tend to lie horizontally on top of the water surface at low coverage, but will tend to assume more upright configurations at high coverage (*see* page 103)

The strong adsorption of such materials at surfaces or interfaces in the form of an orientated *monomolecular layer* (or *monolayer*) is termed *surface activity*. Surface-active materials (or *surfactants*) consist of molecules containing both polar and non-polar parts (*amphiphilic*). Surface activity is a dynamic phenomenon, since the final state of a surface or interface represents a balance between this tendency towards adsorption and the tendency towards complete mixing due to the thermal motion of the molecules.

The tendency for surface-active molecules to pack into an interface favours an expansion of the interface; this must, therefore, be balanced against the tendency for the interface to contract under normal surface tension forces. If π is the expanding pressure (or *surface pressure*) of an adsorbed layer of surfacant, then the surface (or interfacial) tension will be lowered to a value

$$\gamma = \gamma_0 - \pi \tag{4.14}$$

Figure 4.10 shows the effect of lower members of the homologous series of normal fatty alcohols on the surface tension of water. The longer the hydrocarbon chain, the greater is the tendency for the alcohol molecules to adsorb at the air–water surface and, hence, lower the surface tension. A rough generalisation, known as Traube's rule, is that for a particular homologous series of surfactants the concentration required for an equal lowering of surface tension in dilute solution decreases by a factor of about 3 for each additional CH_2 group.

If the interfacial tension between two liquids is reduced to a sufficiently low value on addition of a surfactant, emulsification will readily take place, because only a relatively small increase in the surface free energy of the system is involved. If $\pi \geqslant \gamma_0$, a microemulsion may form (see page 269).

In certain cases – solutions of electrolytes, sugars, etc. – small increases in surface tension due to negative adsorption are noted. Here, because the solute–solvent attractive forces are greater than the solvent–solvent attractive forces, the solute molecules tend to migrate away from the surface into the bulk of the liquid.

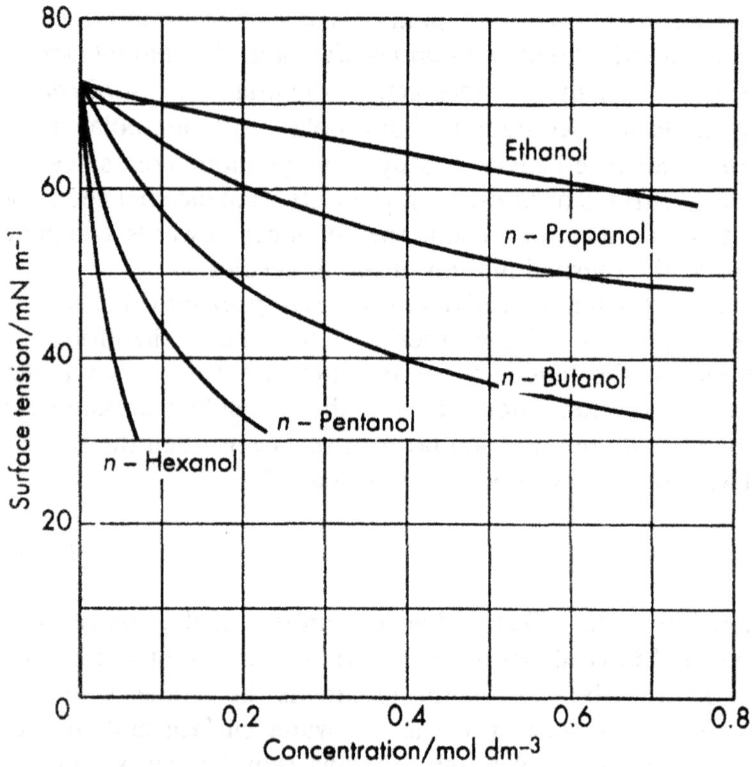

Figure 4.10 Surface tension of aqueous solutions of alcohols at 20°C

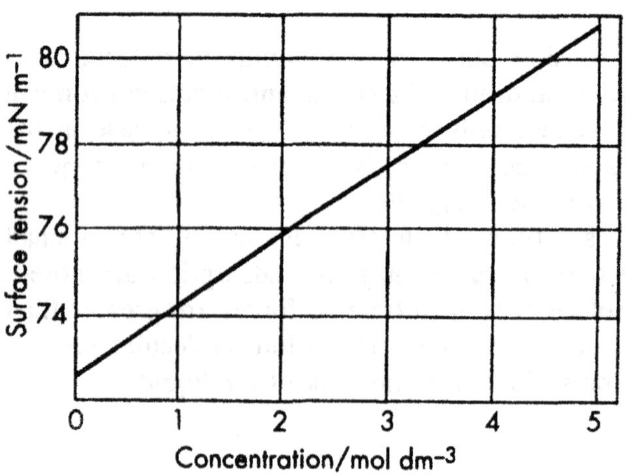

Figure 4.11 Surface tension of aqueous sodium chloride solutions at 20°C

Classification of surfactants[11,14]

The hydrophilic part of the most effective soluble surfactants (e.g. soaps, synthetic detergents and dyestuffs) is often an ionic group. Ions have a strong affinity for water owing to their electrostatic attraction to the water dipoles and are capable of pulling fairly long hydrocarbon chains into solution with them; for example, palmitic acid, which is virtually un-ionised, is insoluble in water, whereas sodium palmitate, which is almost completely ionised, is soluble (especially above its Krafft temperature – see page 93).

It is possible to have non-ionic hydrophilic groups which also exhibit a strong affinity for water; for example, the monomer units in a poly (ethylene oxide) chain each show a modest affinity for water and the sum effect of several of these units in the polymer chain is an overall strong affinity for water.

Surfactants are classified as *anionic*, *cationic*, *non-ionic* or *ampholytic* according to the charge carried by the surface-active part of the molecule. Some common examples are given in Table 4.2. In addition, surfactants are often named in relation to their technological application; hence names such as *detergent*, *wetting agent*, *emulsifier* and *dispersant*.

Anionics are the most widely used surfactants on account of cost and performance. Cationics are expensive, but their germicidal action makes them useful for some applications. An advantage enjoyed by non-ionics is that the lengths of both hydrophilic and hydrophobic groups can be varied.

Rate of adsorption

The formation of an adsorbed surface layer is not an instantaneous process but is governed by the rate of diffusion of the surfactant through the solution to the interface. It might take several seconds for a surfactant solution to attain its equilibrium surface tension, especially if the solution is dilute and the solute molecules are large and unsymmetrical. Much slower ageing effects have been reported, but these are now known to be due to traces of impurities. The time factor in adsorption can be demonstrated by measuring the surface tensions of freshly formed surfaces by a dynamic method; for example, the surface tensions of sodium oleate solutions measured by

Table 4.2 Surface-active agents

Anionic	
Sodium stearate	$CH_3(CH_2)_{16}COO^-Na^+$
Sodium oleate	$CH_3(CH_2)_7CH{=}CH(CH_2)_7COO^-Na^+$
Sodium dodecyl sulphate	$CH_3(CH_2)_{11}SO_4^-Na^+$
Sodium dodecyl benzene sulphonate	$CH_3(CH_2)_{11}.C_6H_4.SO_3^-Na^+$
Cationic	
Dodecylamine hydrochloride	$CH_3(CH_2)_{11}NH_3^+Cl^-$
Hexadecyltrimethyl ammonium bromide	$CH_3(CH_2)_{15}N(CH_3)_3^+Br^-$
Non-ionic	
Polyethylene oxides	e.g. $CH_3(CH_2)_{11}(O.CH_2.CH_2)_6OH$*
Spans (sorbitan esters)	
Tweens (polyoxyethylene sorbitan esters)	
Ampholytic	
Dodecyl betaine	$C_{12}H_{25}N^+\langle^{(CH_3)_2}_{CH_2COO^-}$

*Abbreviated $C_{12}E_6$ to denote hydrocarbon and ethylene oxide chain lengths.

the oscillating jet method approach that of pure water but fall rapidly as the surfaces are allowed to age[46,150].

Thermodynamics of adsorption – Gibbs adsorption equation

The Gibbs adsorption equation enables the extent of adsorption at a liquid surface to be estimated from surface tension data.

The quantitative treatment of surface phenomena involves an important uncertainty. It is convenient to regard the interface between two phases as a mathematical plane, such as SS in Figure 4.12. This approach, however, is unrealistic, especially if an adsorbed film is present. Not only will such a film itself have a certain thickness, but also its presence may influence nearby structure (for example, by dipole–dipole orientation, especially in an aqueous phase) and result in an interfacial region of varying composition with an appreciable thickness in terms of molecular dimensions.

If a mathematical plane is, nevertheless, taken to represent the interface between two phases, adsorption can be described conveni-

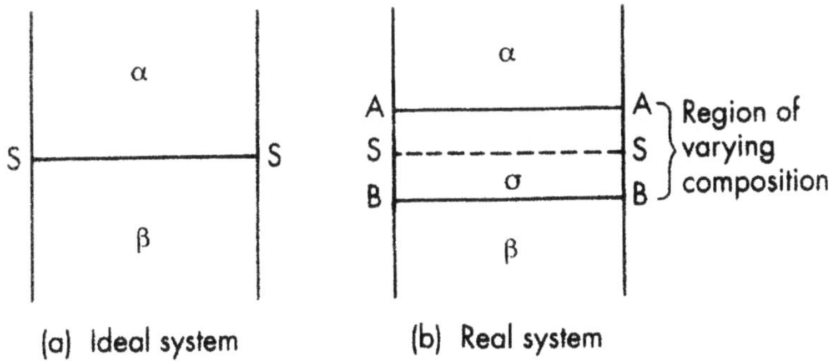

Figure 4.12 Representations of an interface between bulk phases α and β

ently in terms of *surface excess concentrations*. If n_i^σ is the amount of component i in the surface phase σ (Figure 4.12b) in excess of that which would have been in σ had the bulk phases α and β extended to a surface SS with unchanging composition, the surface excess concentration of component i is given by

$$\Gamma_i = \frac{n_i^\sigma}{A} \tag{4.15}$$

where A is the interfacial area. Γ_i may be positive or negative, and its magnitude clearly depends on the location of SS, which (as illustrated in the following derivation) must be chosen somewhat arbitrarily.

The total thermodynamic energy of a system is given by the expression

$$U = TS - pV + \Sigma\mu_i n_i$$

The corresponding expression for the thermodynamic energy of a surface phase σ is

$$U^\sigma = TS^\sigma - pV^\sigma + \gamma A + \Sigma\mu_i n_i^\sigma \tag{4.16}$$

(The pV^σ and γA terms have opposite signs, since pressure is an expanding force and surface tension is a contracting force. A superscript is not necessary for T, p and the chemical potential terms, since these must have uniform values throughout for a heterogeneous system to be in equilibrium.) Differentiating equation (4.16) generally

$$\mathrm{d}U^\sigma = T\mathrm{d}S^\sigma + S^\sigma\mathrm{d}T - p\mathrm{d}V^\sigma - V^\sigma\mathrm{d}p + \gamma\mathrm{d}A + A\mathrm{d}\gamma + \Sigma\mu_i\mathrm{d}n_i^\sigma + \Sigma n_i^\sigma\mathrm{d}\mu_i \tag{4.17}$$

From the first and second laws of thermodynamics,

$$dU = TdS - pdV + \Sigma\mu_i dn_i$$

or, for a surface phase,

$$dU^\sigma = TdS^\sigma - pdV^\sigma + \gamma dA + \Sigma\mu_i dn_i^\sigma \tag{4.18}$$

Subtracting equation (4.18) from equation (4.17),

$$S^\sigma dT - V^\sigma dp + Ad\gamma + \Sigma n_i^\sigma \, d\mu_i = 0$$

Therefore, at constant temperature and pressure

$$d\gamma = -\Sigma \frac{n_i^\sigma}{A} d\mu_i = -\Sigma\Gamma_i d\mu_i \tag{4.19}$$

For a simple two-component solution (i.e. consisting of a solvent and a single solute) equation (4.19) becomes

$$d\gamma = -\Gamma_A d\mu_A - \Gamma_B d\mu_B$$

As explained above, surface excess concentrations are defined relative to an arbitrarily chosen dividing surface. A convenient (and seemingly realistic) choice of location of this surface for a binary solution is that at which the surface excess concentration of the solvent (Γ_A) is zero. The above expression then simplifies to

$$d\gamma = -\Gamma_B d\mu_B$$

Since chemical potential changes are related to relative activities by

$$\mu_B = \mu_B^\ominus + RT \ln a_B$$

then $d\mu_B = RT \, d \ln a_B$

Therefore $$\Gamma_B = -\frac{1}{RT} \cdot \frac{d\gamma}{d \ln a_B} = -\frac{a_B}{RT} \cdot \frac{d\gamma}{da_B} \tag{4.20}$$

or, for dilute solutions,

$$\Gamma_B = -\frac{c_B}{RT} \cdot \frac{d\gamma}{dc_B} \tag{4.21}$$

which is the form in which the Gibbs equation is usually quoted.

The Gibbs equation in this form could be applied to a solution of a non-ionic surfactant. For a solution of an ionic surfactant in the absence of any other electrolyte, Haydon and co-workers[3,151] have argued that equations (4.20) and (4.21) should be modified to allow for the fact that both the anions and the cations of the surfactant will adsorb at the solution surface in order to maintain local electrical neutrality (even though not all of these ions are surface-active in the amphiphilic sense). For a solution of a 1 : 1 ionic surfactant a factor of 2 is required to allow for this simultaneous adsorption of cations and anions, and equation (4.21) must be modified to

$$\Gamma_B = \frac{-c_B}{2RT} \cdot \frac{d\gamma}{dc_B} \tag{4.22}$$

In the presence of excess inert electrolyte, however, an electrical shielding effect will operate and equation (4.21) will apply.

Experimental verification of the Gibbs equation

The general form of the Gibbs equation ($d\gamma = -\Sigma\,\Gamma_i d\mu_i$) is fundamental to all adsorption processes. However, experimental verification of the equation derived for simple systems is of interest in view of the postulation which was made concerning the location of the boundary surface.

McBain and Swain[152] succeeded in verifying the validity of the Gibbs equation by means of a very direct and ingenious experiment. Surface layers of about 0.1 mm thickness were shaved off solutions of surface-active materials, such as phenol and hydrocinnamic acid, contained in a long rectangular trough, by means of a rapidly moving microtome blade. The material collected was analysed and experimental surface excess concentrations were calculated. These compared well with the corresponding surface excess concentrations calculated from surface tension data.

Surface concentrations have also been successfully measured[47] by labelling the solute with a β-emitting radioactive isotope (e.g. ^{3}H, ^{14}C, ^{35}S or ^{45}Ca) and measuring the radiation picked up by a Geiger counter placed immediately above the surface of the solution. As β-rays are rapidly attenuated in the solution, the measured radiation corresponds to the surface region plus only a thin layer of bulk region. Direct measurement of surface concentrations is particularly useful when there is more than one surface-active species or unavoidable surface-active impurity present. In such cases surface tension measurements would probably be ambiguous.

Association colloids – micelle formation[48]

Physical properties of surfactant solutions

Solutions of highly surface-active materials exhibit unusual physical properties. In dilute solution the surfactant acts as a normal solute (and in the case of ionic surfactants, normal electrolyte behaviour is observed). At fairly well defined concentrations, however, abrupt changes in several physical properties, such as osmotic pressure, turbidity, electrical conductance and surface tension, take place (see Figure 4.13). The rate at which osmotic pressure increases with concentration becomes abnormally low and the rate of increase of turbidity with concentration is much enhanced, which suggests that considerable association is taking place. The conductance of ionic surfactant solutions, however, remains relatively high, which shows that ionic dissociation is still in force.

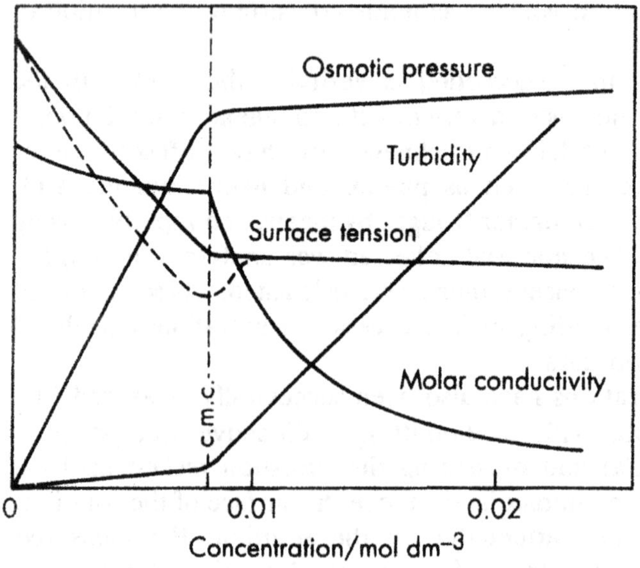

Figure 4.13 Physical properties of sodium dodecyl sulphate solutions at 25°C

McBain pointed out that this seemingly anomalous behaviour could be explained in terms of organised aggregates, or *micelles*, of the surfactant ions in which the lipophilic hydrocarbon chains are orientated towards the interior of the micelle, leaving the hydrophilic groups in contact with the aqueous medium. The concentration above which micelle formation becomes appreciable is termed the *critical micelle concentration* (c.m.c.).

Micellisation is, therefore, an alternative mechanism to adsorption by which the interfacial energy of a surfactant solution might decrease.

When one considers the energetics of micellisation in terms of the hydrocarbon chains of the surfactant molecules, the following factors are among those which must be taken into account:

1. The intermolecular attractions between the hydrocarbon chains in the interior of the micelle represent an energetically favourable situation; but it is not one which is significantly more favourable than that which results from the alternative hydrocarbon–water attraction in the case of single dissolved surfactant molecules. Comparison of the surface tension of a typical hydrocarbon oil with the dispersion component of the surface tension of water (as discussed on page 67) illustrates this point.
2. Micellisation permits strong water–water interaction (hydrogen bonding) which would otherwise be prevented if the surfactant was in solution as single molecules wedged between the solvent water molecules. This is a most important factor in micelle formation and also of course, in any adsorption process at an aqueous interface. It is often referred to as the *hydrophobic effect*[49].

Experimental study of micelles

Critical micelle concentrations can be determined by measuring any micelle-influenced physical property as a function of surfactant concentration. In practice, surface tension, electrical conductivity and dye solubilisation measurements (see Figure 4.13 and page 90) are the most popular. The choice of physical property will slightly influence the measured c.m.c., as will the procedure adopted to determine the point of discontinuity.

Information concerning the sizes and shapes of micelles can be obtained from dynamic light scattering (page 61), ultracentrifugation (page 31), viscosity and low-angle X-ray scattering.

Factors affecting critical micelle concentrations

1. Increasing the hydrophobic part of the surfactant molecules favours micelle formation (see Table 4.3). In aqueous medium, the c.m.c. of ionic surfactants is approximately halved by the addition of each CH_2 group. For non-ionic surfactants this effect is usually even more pronounced. This trend usually continues up to about the C_{16} member. Above the C_{18} member the c.m.c. tends to be approximately constant. This is probably the result of coiling of the long hydrocarbon chains in the water phase[50].

Table 4.3 Critical micelle concentrations for a homologous series of sodium alkyl sulphates in water at 40°C[11]

Number of carbon atoms	8	10	12	14	16	18
c.m.c./10^{-3} mol dm^{-3}	140	33	8.6	2.2	0.58	0.23

2. Micelle formation is opposed by thermal agitation and c.m.c.'s would thus be expected to increase with increasing temperature. This is usually, but not always, the case, as discussed on page 93.
3. With ionic micelles, the addition of simple electrolyte reduces the repulsion between the charged groups at the surface of the micelle by the screening action of the added ions (see Chapter 7). The c.m.c. is, therefore, lowered, as illustrated in Table 4.4.

Table 4.4 Critical micelle concentrations of sodium dodecyl sulphate in aqueous sodium chloride solutions at 25°C[11]

c. (NaCl)/mol dm^{-3}	0	0.01	0.03	0.1	0.3
c.m.c./10^{-3} mol dm^{-3}	8.1	5.6	3.1	1.5	0.7

4. The addition of organic molecules can affect c.m.c.'s in a variety of ways. The most pronounced changes are effected by those molecules (e.g. medium chain-length alcohols, see page 89) which can be incorporated into the outer regions of the micelle. There they can reduce electrostatic repulsion and steric hindrance, thus

lowering the c.m.c. Micelles containing more than one surfactant often form readily with a c.m.c. lower than any of the c.m.c.'s of the pure constituents.

Organic molecules may influence c.m.c.'s at higher additive concentrations by virtue of their influence on water structuring. Sugars are structure-makers and as such cause a lowering of c.m.c., whereas urea and formamide are structure-breakers and their addition causes an increase in c.m.c.

Structure of micelles

Micellar theory has developed in a somewhat uncertain fashion and is still in many respects open to discussion. Possible micelle structures include the spherical, laminar and cylindrical arrangements illustrated schematically in Figure 4.14. Living cells can be considered as micellar-type arrangements with a vesicular structure.

Typically, micelles tend to be approximately spherical over a fairly wide range of concentration above the c.m.c., but often there are marked transitions to larger, non-spherical liquid-crystal structures at high concentrations. Systems containing spherical micelles tend to have low viscosities, whereas liquid-crystal phases tend to have high viscosities. The free energies of transition between micellar phases tend to be small and, consequently, the phase diagrams for these systems tend to be quite complicated and sensitive to additives.

Some of the experimental evidence favouring the existence of spherical, liquid-like micelles is summarised, as follows:

1. Critical micelle concentrations depend almost entirely on the nature of the lyophobic part of the surfactant. If micelle structure involved some kind of crystal lattice arrangement, the nature of the lyophilic head group would also be expected to be important.
2. The micelles of a given surfactant are approximately monodispersed and their size depends predominantly on the nature of the lyophobic part of the surfactant molecules. One would expect the radius of spherical micelles to be slightly less than the length of the constituent units; otherwise the hydrocarbon chains would be considerably buckled or the micelle would have either a hole or ionic groups in the centre. The radii of micelles calculated from diffusion and light-scattering data support this expectation. For straight-chain ionic surfactants the number of monomer units per

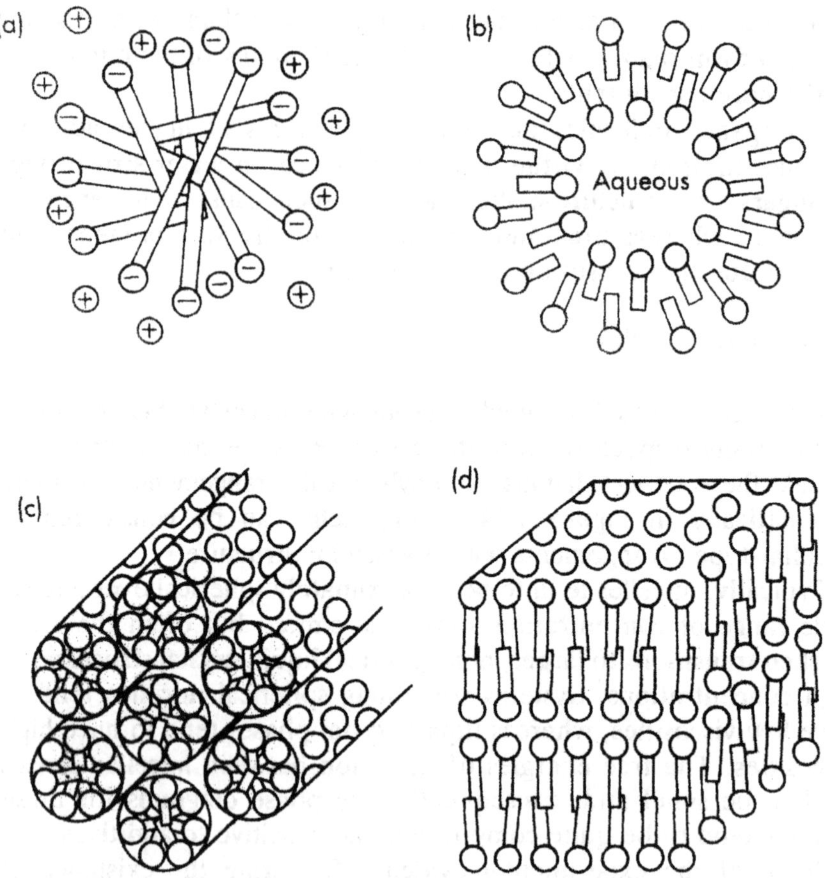

Figure 4.14 Micellar structures. (a) Spherical (anionic) micelle. This is the usual shape at surfactant concentrations below about 40 per cent. (b) Spherical vesicle bilayer structure (*see also Figure 4.28*), which is representative of the living cell. (c) and (d) Hexagonal and lamellar phases formed from cylindrical and laminar micelles, respectively. These, and other structures, exist in highly concentrated surfactant solutions

micelle, *m*, and the number of carbon atoms per hydrocarbon chain, *n*, are approximately related as follows:

n	12	14	16	18
m	33	46	60	78

Laminar and cylindrical models, in contrast, provide no satisfactory mechanism by which the size of the micelles might be limited.

3. For diffusion reasons, solubilisation (see next section) would not take place readily if the micelle were solid.

As mentioned above, the length of the surfactant's hydrocarbon chain will dictate the radius of a spherical micelle. This in turn determines the spacing of the outer polar groups. On this basis, for example, a dodecyl sulphate micelle surface would be expected to be approximately one-third sulphate groups and two-thirds hydrocarbon. The results of neutron scattering studies are consistent with this expectation. In an ionic micelle, the tendency of this hydrocarbon–water interfacial area to contract is balanced by head-group repulsion. Addition of electrolyte reduces this head-group repulsion, thus favouring an area per head-group that is smaller than the geometric optimum for a spherical micelle. Under such conditions, the micelle is likely to distort to a non-spherical shape.

There is evidence from nuclear magnetic resonance spectroscopy and partial molar volume measurements[153–154] which points to the possible existence of bound water in the micelle interior in the region of the first few CH_2 groups in from the polar head groups. The hydrocarbon interior of the micelle may, therefore, be considered in terms of an outer region which may be penetrated by water and an inner region from which water is excluded.

Solubilisation[51]

Surfactant solutions above the c.m.c. can solubilise otherwise insoluble organic material by incorporating it into the interior of the micelles; for example, the dye xylenol orange dissolves only sparingly in pure water but gives a deep red solution with sodium dodecyl sulphate present above its c.m.c.

The balance of electrostatic and hydrophobic interactions can be such as to cause the locus of solubilisation to be anywhere in the micelle from close to the surface to the inner core.

Solubilisation is of practical importance in the formulation of pharmaceutical and other products containing water-insoluble ingredients[51], detergency, where it plays a major role in the removal of oily soil (pages 166–176), emulsion polymerisation (page 17) and micellar catalysis of organic reactions[52].

In micellar catalysis, reactant must be solubilised at a location near to the micelle surface where it is accessible to reagent in the aqueous phase. The strong electrostatic interactions which are likely at this location may influence the nature of the transition state and/or reactant concentration; for example, cationic micelles may catalyse reaction between a nucleophilic anion and a neutral solubilised substrate.

Surface behaviour

Figure 4.13 illustrates how a highly surface-active material such as sodium dodecyl sulphate lowers the surface tension of water quite appreciably even at low concentrations. The discontinuity in the γ-composition curve is identified with the c.m.c., beyond which there is an additional mechanism for keeping hydrocarbon chains away from water surfaces – i.e. by locating them in the interior of the micelles. Since the micelles themselves are not surface-active, the surface tension remains approximately constant beyond the c.m.c. The minimum in the γ-composition curve, shown by the dashed curve, is typical of measurements which have been made on surfactant solutions and in apparent violation of the Gibbs equation, since it suggests desorption over the small concentration range where $\mathrm{d}\gamma/\mathrm{d}c$ is positive. This anomaly is attributed to traces of impurity such as dodecanol, which is surface-adsorbed below the c.m.c. but solubilised by the micelles beyond the c.m.c. With sufficient purification the minimum in the γ-composition curve can be removed. Beyond the c.m.c., where $\mathrm{d}\gamma/\mathrm{d}c \sim 0$, application of the Gibbs equation might suggest almost zero adsorption; however, $\mathrm{d}\gamma/\mathrm{d}a$, where a represents the activity of single surfactant species, is still appreciably negative, a changing little above the c.m.c.

Conductance

Micelle formation affects the conductance of ionic surfactant solutions for the following reasons:

1. The total viscous drag on the surfactant molecules is reduced on aggregation.
2. Counter-ions become kinetically a part of the micelle, owing to its high surface charge (see Chapter 7), thus reducing the number of counter-ions available for carrying the current and also lowering the net charge of the micelles. Typically, 50 to 70% of the counter-

ions are held in the Stern layer; even so, the zeta potential of an ionic micelle is usually high.

3. The retarding influence of the ionic atmospheres of unattached counter-ions on the migration of the surfactant ions is greatly increased on aggregation.

The last two factors, which cause the molar conductivity to decrease with concentration beyond the c.m.c., normally outweigh the first factor, which has the reverse effect (see Figure 4.13). When conductance measurements are made at very high field strengths the ionic atmospheres cannot re-form quickly enough (Wien effect) and some of the bound counter-ions are set free. It is interesting to note that under these conditions the molar conductivity increases with concentration beyond the c.m.c.

Sharpness of critical micelle concentrations

There are two current theories relating to the abruptness with which micellisation takes place above a certain critical concentration[53,155].

The first of these theories applies the law of mass action to the equilibrium between unassociated molecules or ions and micelles, as illustrated by the following simplified calculation for the micellisation of non-ionic surfactants. If c is the stoichiometric concentration of the solution, x is the fraction of monomer units aggregated and m is the number of monomer units per micelle,

$$\underset{c(1-x)}{mX} = \underset{cx/m}{(X)_m}$$

Therefore, applying the law of mass action,

$$K = \frac{cx/m}{[c(1-x)]^m} \tag{4.23}$$

For moderately large values of m, this expression requires that x should remain very small up to a certain value of c and increase rapidly thereafter. The sharpness of the discontinuity will depend on the value of m ($m = \infty$ gives a perfect discontinuity). If this treatment is modified to account for the counter-ions associated with an ionic micelle, then an even more abrupt discontinuity than the above is predicted.

The alternative approach is to treat micellisation as a simple phase separation of surfactant in an associated form, with the unassociated surfactant concentration remaining practically constant above the c.m.c.

In the build-up from surfactant monomers to micelles, the existence, albeit transitorily, of intermediate levels of aggregation is to be expected. Close examination of experimental evidence suggests that there may be some smoothness of property change at the c.m.c., but that sub-micelle aggregates exist only in trace amounts.

Energetics of micellisation

Since the equilibrium constant, $^{\ominus}K$, in equation (4.23) and the standard free energy change, $\Delta G^{\ominus}$, for the micellisation of 1 mole of surfactant are related by

$$\Delta G^{\ominus} = -\frac{RT}{m} \ln K \tag{4.24}$$

then, substituting into equation (4.23),

$$\Delta G^{\ominus} = -\frac{RT}{m} \ln\left(\frac{cx}{m}\right) + RT \ln[c(1-x)] \tag{4.25}$$

At the c.m.c. (neglecting pre-c.m.c. association and, therefore, reverting to the phase separation model), $x = 0$ and

$$\Delta G^{\ominus} = RT \ln \text{ (c.m.c.)} \tag{4.26}$$

Therefore, $$\Delta S^{\ominus} = -\frac{\mathrm{d}(\Delta G^{\ominus})}{\mathrm{d}T}$$

$$= -RT\frac{\mathrm{d} \ln \text{ (c.m.c)}}{\mathrm{dT}} - R \ln \text{ (c.m.c.)} \tag{4.27}$$

and $\Delta H^{\ominus} = \Delta G^{\ominus} + T\Delta S^{\ominus}$

$$= -RT^2 \frac{d\ln \text{ (c.m.c.)}}{\mathrm{d}T} \tag{4.28}$$

In general, micellisation is an exothermic process and the c.m.c. increases with increasing temperature (see page 86). This, however, is not universally the case: for example, the c.m.c. of sodium dodecyl sulphate in water shows a shallow minimum between about 20°C and 25°C. At lower temperature the enthalpy of micellisation given from equation (4.28) is positive (endothermic), and micellisation is entirely entropy-directed.

The cause of a positive entropy of micellisation is not entirely clear. A decrease in the amount of water structure as a result of micellisation may make some contribution. A more likely contribution, however, involves the configuration of the hydrocarbon chains, which probably have considerably more freedom of movement in the interior of the micelle than when in contact with the aqueous medium.

The Krafft phenomenon

Micelle-forming surfactants exhibit another unusual phenomenon in that their solubilities show a rapid increase above a certain temperature, known as the *Krafft point*. The explanation of this behaviour arises from the fact that unassociated surfactant has a limited solubility, whereas the micelles are highly soluble. Below the Krafft temperature the solubility of the surfactant is insufficient for micellisation. As the temperature is raised, the solubility slowly increases until, at the Krafft temperature, the c.m.c. is reached. A relatively large amount of surfactant can now be dispersed in the form of micelles, so that a large increase in solubility is observed.

Table 4.5 Krafft temperatures for sodium alkyl sulphates in water

Number of carbon atoms	10	12	14	16	18
Krafft temperature/°C	8	16	30	45	56

Spreading

Adhesion and cohesion

The *work of adhesion* between two immiscible liquids is equal to the work required to separate unit area of the liquid–liquid interface and form two separate liquid–air interfaces (Figure 4.15a), and is given by the Dupré equation

$$W_a = \gamma_A + \gamma_B - \gamma_{AB} \tag{4.29}$$

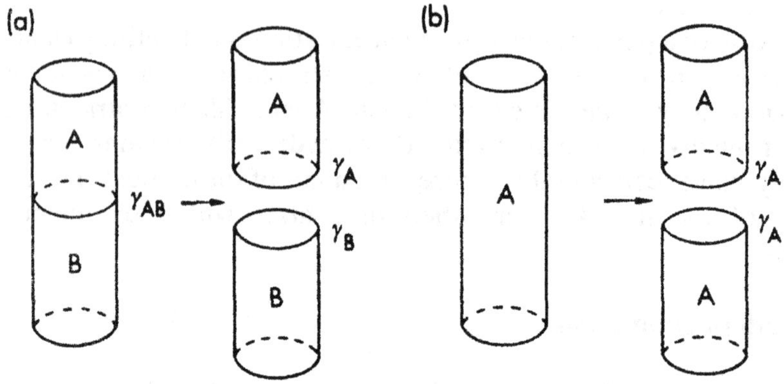

Figure 4.15 Work of adhesion (a) and of cohesion (b)

The *work of cohesion* for a single liquid corresponds to the work required to pull apart a column of liquid of unit cross-sectional area (Figure 4.15b) – i.e.

$$W_c = 2\gamma_A \tag{4.30}$$

Spreading of one liquid on another

When a drop of an insoluble oil is placed on a clean water surface, it may behave in one of three ways:

1. Remain as a lens, as in Figure 4.16 (non-spreading).
2. Spread as a thin film, which may show interference colours, until it is uniformly distributed over the surface as a 'duplex' film. (A duplex film is a film which is thick enough for the two interfaces – i.e. liquid–film and film–air – to be independent and possess characteristic surface tensions.)
3. Spread as a monolayer, leaving excess oil as lenses in equilibrium, as in Figure 4.17.

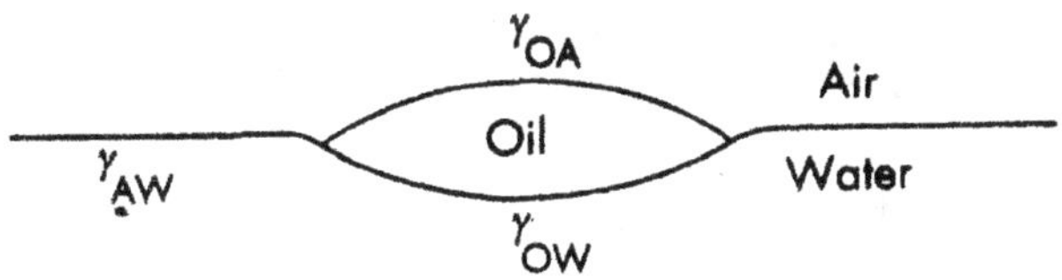

Figure 4.16 A drop of non-spreading oil on a water surface

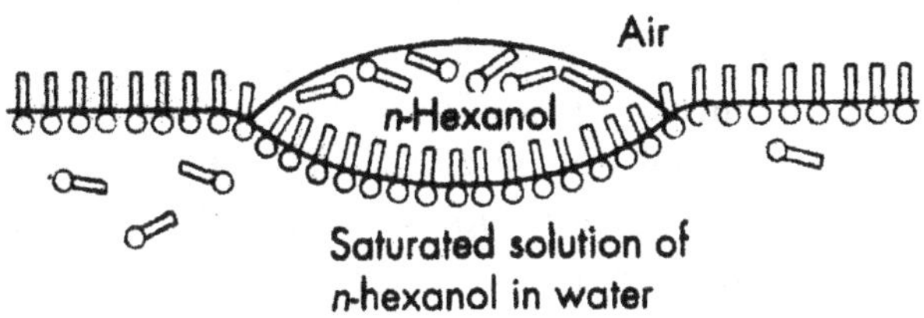

Figure 4.17 Spreading of *n*-hexanol on a water surface

If the area occupied by the oil drop shown in Figure 4.16 is increased by dA, the change in the surface free energy of the system will be approximately $(\gamma_{OA} + \gamma_{OW} - \gamma_{WA})$ dA. If this quantity is negative, the process of spreading will take place spontaneously.

Harkins defined the term *initial spreading coefficient* (for the case of oil on water) as

$$S = \gamma_{WA} - (\gamma_{OA} + \gamma_{OW}) \tag{4.31}$$

where the various interfacial tensions are measured before mutual saturation of the liquids in question has occurred. The condition for initial spreading is therefore that S be positive or zero (Table 4.6).

Table 4.6 Initial spreading coefficients (in mN m^{-1}) for liquids on water at 20°C[54] (*By courtesy of Academic Press Inc.*)

Liquid	$\gamma_{WA} - (\gamma_{OA} + \gamma_{OW}) = S$	*Conclusion*
n-Hexadecane	72.8−(30.0+52.1) = −9.3	will not spread on water
n-Octane	72.8−(21.8+50.8) = +0.2	will just spread on pure water
n-Octanol	72.8−(27.5+ 8.5) = +36.8	will spread against contamination

Substituting in the Dupré equation, the spreading coefficient can be related to the work of adhesion and cohesion

$$S = W_{OW} - 2\gamma_{OA} = W_{OW} - W_{Oil} \tag{4.32}$$

i.e. spreading occurs when the oil adheres to the water more strongly than it coheres to itself.

Impurities in the oil phase (e.g. oleic acid in hexadecane) can reduce γ_{OW} sufficiently to make S positive. Impurities in the aqueous phase normally reduce S, since γ_{WA} is lowered more than γ_{OW} by the impurity, especially if γ_{OW} is already low. Therefore, *n*-octane will spread on a clean water surface but not on a contaminated surface.

The initial spreading coefficient does not consider the mutual saturation of one liquid with another: for example, when benzene is spread on water,

$$S_{init.} / \text{mN m}^{-1} = 72.8 - (28.9 + 35.0) = +8.9$$

but when the benzene and water have had time to become mutually saturated, γ_{WA} is reduced to 62.4 mN m^{-1} and γ_{OW} to 28.8 mN m^{-1}, so that

$$S_{final} / \text{mN m}^{-1} = 62.4 - (28.8 + 35.0) = -1.4$$

The final state of the interface is now just unfavaourable for spreading. This causes the initial spreading to be stopped, and can even result in the film retracting slightly to form very flat lenses, the rest of the water surface being covered by a monolayer of benzene.

Similar considerations apply to the spreading of a liquid such as *n*-hexanol on water (Figure 4.17):

$$S_{init.} / \text{mN m}^{-1} = 72.8 - (24.8 + 6.8) = +41.2$$

$$S_{final} / \text{mNm}^{-1} = 28.5 - (24.7 + 6.8) = -3.0$$

Monomolecular films[54–57]

Many insoluble substances, such as long-chain fatty acids and alcohols, can (with the aid of suitable solvents) be spread on to a water surface, and if space permits will form a surface film one molecule in thickness with the hydrophilic $-COOH$ or $-OH$ groups orientated towards the water phase and the hydrophobic hydrocarbon chains orientated away from the water phase.

These insoluble monomolecular films, or monolayers, represent an extreme case in adsorption at liquid surfaces, as all the molecules in question are concentrated in one molecular layer at the interface. In this respect they lend themselves to direct study. In contrast to monolayers which are formed by adsorption from solution, the surface concentrations of insoluble films are known directly from the amount of material spread and the area of the surface, recourse to the Gibbs equation being unnecessary.

The molecules in a monomolecular film, especially at high surface concentrations, are often arranged in a simple manner, and much can be learned about the size, shape and orientation of the individual molecules by studying various properties of the monolayer. Monomolecular films can exist in different, two-dimensional physical states, depending mainly on the magnitude of the lateral adhesive forces between the film molecules, in much the same way as three-dimensional matter.

Experimental techniques for studying insoluble monolayers

Surface pressure

The surface pressure of a monolayer is the lowering of surface tension due to the monolayer – i.e. it is the expanding pressure exerted by the monolayer which opposes the normal contracting tension of the clean interface, or

$$\pi = \gamma_0 - \gamma \tag{4.14}$$

where γ_0 is the tension of clean interface and γ is the tension of interface plus monolayer.

The variation of surface pressure with the area available to the spread material is represented by a π–A (force–area) curve. With a little imagination, π–A curves can be regarded as the two-dimensional equivalent of the p–V curves for three dimensional matter. (N.B. For a 1 nm thick film, a surface pressure of 1 mN m^{-1} is equivalent to a bulk pressure of 10^6 N m^{-2}, or $\sim$ 10 atm.)

The Langmuir–Adam surface balance (or trough) uses a technique for containing and manoeuvring insoluble monolayers between barriers for the direct determination of π–A curves. The film is contained (Figure 4.18) between a movable barrier and a float attached to a torsion-wire arrangement. The surface pressure of the film is measured directly in terms of the horizontal force which it exerts on the float and the area of the film is varied by means of the movable barrier.

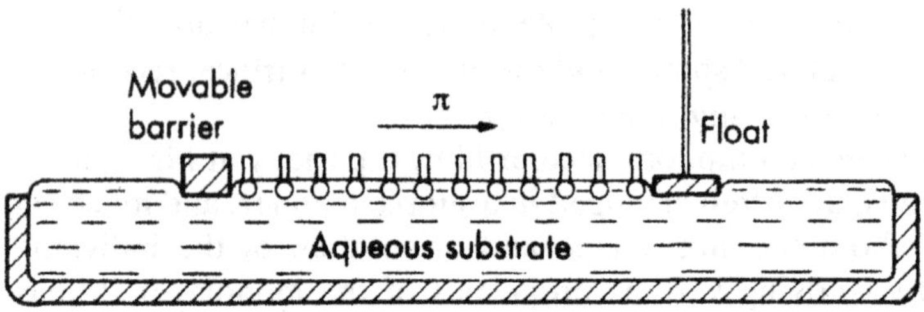

Figure 4.18 The principle of the Langmuir–Adam surface balance

The film must be contained entirely between the barrier and the float without any leakage. To achieve this, the trough walls, barrier and float must be hydrophobic and the liquid level must be slightly above the brim of the trough. Teflon troughs and accessories are ideal in this respect. Silica or glass apparatus which has been made hydrophobic with a light coating of purified paraffin wax or silicone waterproofing material is also satisfactory. Waxed threads prevent leakage past the ends of the float.

Before spreading, the surface must be carefully cleaned and freed from contamination. Redistilled water should be used. The liquid surface in front of and behind the float can be swept clean by moving barriers towards the float and sucking away any surface impurities with a capillary joined to a water pump.

To achieve uniform spreading, the material in question is normally predissolved in a solvent such as petroleum ether to give a *c.* 0.1 per cent solution. A total of *c.* 0.01 cm^3 of this spreading solution is ejected in small amounts from a micrometer syringe burette at various points on the liquid surface. The spreading solvent evaporates away, leaving a uniformly spread film. Benzene, although frequently used as a spreading solvent, is not entirely suitable, owing to its slight solubility in water and long residence time at the interface (see page 96).

Figure 4.19 A surface balance (*By courtesy of Unilever Research Laboratory, Port Sunlight*)

The surface pressure of the film is determined by measuring the force which must be applied via a torsion wire to maintain the float at a fixed position on the surface (located optically) and dividing by the length of the float. For precise work, the surface balance is enclosed in an air thermostat and operated by remote control. With a good modern instrument, surface pressures can be measured with an accuracy of 0.01 mN m^{-1}.

Surface pressures can also be determined indirectly by measuring the lowering of the surface tension of the underlying liquid (or substrate) caused by the film. For example, the float can be replaced by a Wilhelmy plate set-up (Figure 4.5b). This method is at least as accurate as the Langmuir–Adam surface balance and is particularly useful for studying films at oil–water interfaces[156]. The obvious advantage of this method is that it is not necessary to maintain a clean surface behind a float. The arrangement can be simplified even further by doing away with the movable barrier as well and regulating the area of the film by the amount of material spread.

Surface film potential

In heterogeneous systems, potential differences exist across the various phase boundaries. The *surface film potential*, ΔV, due to a monolayer is the change in the potential difference between the bulk substrate liquid and a probe placed above the liquid which results from the presence of the monolayer. Surface film potentials can be measured by air-electrode and vibrating-plate methods.

The air electrode consists of an insulated metal wire with its tip 1–2 mm above the liquid surface and with polonium deposited on the tip to make the air gap conducting.

In the vibrating-plate method, a small gold or gold-plated disc is mounted about 0.5 mm above the surface. The vibration of the disc (~ 200 Hz) produces a corresponding variation in the capacity across the air gap, an alternating current being thus set up, the magnitude of which depends on the potential difference across the gap. This method is more accurate than the air-electrode method, being capable of measuring ΔV to 0.1 mV, but is subject to malfunctioning. It can also be used at oil–water interfaces.

Surface film potential measurements can yield useful, if not absolute, information about the orientation of the film molecules. Treating the film as a parallel plate condenser leads to the approximate expression

$$\Delta V = \frac{n\mu \cos\theta}{\epsilon} \tag{4.33}$$

where n is the number of film molecules per unit area, μ is the dipole moment of film molecules, θ is the angle of inclination of dipoles to the normal and ϵ is the permittivity of the film (see page 179).

Surface film potential measurements are also used to investigate the homogeneity or otherwise of the surface. If there are two surface phases present, the surface film potential will fluctuate wildly as the probe is moved across the surface or as one blows gently over the surface.

Surface rheology[58,157]

Surface viscosity is the change in the viscosity of the surface layer brought about by the monomolecular film. Monolayers in different physical states can readily be distinguished by surface viscosity measurements.

A qualitative idea of the surface viscosity is given by noting the ease with which talc can be blown about the surface. Most insoluble films have surface viscosities of *c.* 10^{-6} kg s^{-1} to 10^{-3} kg s^{-1} (for films 10^{-9} m thick this is equivalent to a bulk viscosity range of *c.* 10^3 kg m^{-1} s^{-1} to 10^6 kg m^{-1} s^{-1}). These films can be studied by means of a damped oscillation method (Figure 4.20). For a vane of length l and a disc with a moment of inertia I,

$$\eta_s = \frac{9.2I}{l^2}\left(\frac{\lambda}{t} - \frac{\lambda_0}{t_0}\right) \tag{4.34}$$

where λ = $\log_{10}$ of the ratio of successive amplitudes of damped oscillation, t is the period of oscillation and the subscripts '$_0$' refer to a clean surface.

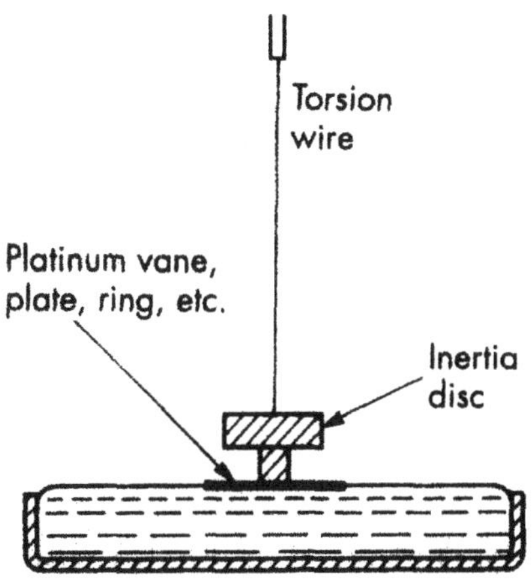

Figure 4.20 Damped oscillation method for measuring surface viscosities

Many insoluble films, particularly those containing protein, exhibit viscoelastic behaviour (see Chapter 9). A surface rheometer has been designed[158] for studying film creep under a constant shearing stress. A platinum ring suspended from a torsion wire is kept under constant torsional stress and its rotation in the plane of the interface is measured as a function of time.

The study of surface rheology is useful in connection with the stability of emulsions and foams (Chapter 10) and the effectiveness of lubricants, adhesives, etc.

Electron micrographs of monolayers

Insoluble surface films can be studied by electron microscopy. The films are transferred from the substrate on to a collodion support and shadow-cast by a beam of metal atoms directed at an angle α (about 15°) to the surface (Figure 4.21). If the width x of the uncoated region is measured, the thickness of the film, $x \tan \alpha$, can be calculated; for example, a n-$C_{36}H_{73}COOH$ film has been shown to be about 5 nm thick – i.e. consistent with a vertically orientated monomolecular layer. The technique has also been used for following the state of the surface as a film is compressed.

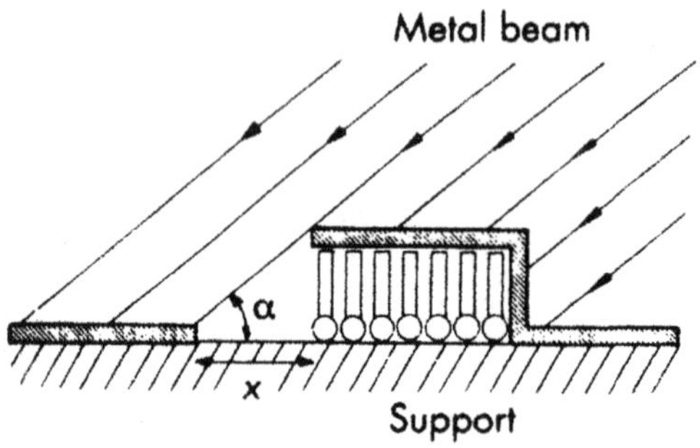

Figure 4.21

The physical states of monomolecular films

Two-dimensional monolayers can exist in different physical states which bear some resemblance to the solid, liquid and gaseous states in three-dimensional matter. Surface films are best classified according to the lateral adhesion between the film molecules, including end-groups. Factors such as ionisation (and, hence, the pH of the substrate) and temperature play an important part in determining the nature of the film. Monolayers can be roughly classified as:

1. *Condensed* (solid) films, in which the molecules are closely packed and steeply orientated towards the surface.
2. Films which are still coherent but occupy a much larger area than condensed films. They have no real three-dimensional equivalent, since they act as highly compressible liquids. A number of distinct types of these *expanded* films have been recognised[21], the most important being the *liquid-expanded state*, but these will not be considered in detail.
3. *Gaseous* or *vapour* films, in which the molecules are separate and move about the surface independently, the surface pressure being exerted on the barriers containing the film by a series of collisions.

Gaseous films

The principal requirements for an ideal gaseous film are that the constituent molecules must be of negligible size with no lateral adhesion between them. Such a film would obey an ideal two-dimensional gas equation, $\pi A = kT$, i.e. the π–A curve would be a rectangular hyperbola. This ideal state of affairs is, of course, unrealisable but is approximated to by a number of insoluble films, especially at high areas and low surface pressures. Monolayers of soluble material are normally gaseous. If a surfactant solution is sufficiently dilute to allow solute–solute interactions at the surface to be neglected, the lowering of surface tension will be approximately linear with concentration – i.e.

$$\gamma = \gamma_0 - bc \text{ (where } b \text{ is a constant)}$$

Therefore

$$\pi = bc$$

and $\mathrm{d}\gamma/\mathrm{d}c = -b$

Substituting in the Gibbs equation

$$\left(\Gamma = \frac{-c}{kT}\frac{\mathrm{d}\gamma}{\mathrm{d}c}\right)$$

gives $\Gamma = \dfrac{1}{A} = \dfrac{\pi}{kT}$

i.e. $\pi A = kT$ (4.35)

where A is the average area per molecule.

An example of a gaseous film is that of cetyl trimethyl ammonium bromide (Figure 4.22). The molecules in the film are ionised to $C_{16}H_{33}N(CH_3)_3{}^+$ and, therefore, repel one another in the aqueous phase, so that π is relatively large at all points on the π–A curve. Film pressures are greater for a given area at the oil–water interface than at the air–water interface, because the oil penetrates between the hydrocarbon chains of the film molecules and removes most of the inter-chain attraction. The π–A curve at the air–water interface approximates to $\pi A = kT$, presumably because inter-chain attractions and electrical repulsions are of the same order of magnitude, whereas at the oil–water interface $\pi A > kT$, because the electrical repulsion between the film molecules outweighs the inter-chain attraction. Fatty acids and alcohols of chain-length C_{12} and less give imperfect gaseous films when spread on water at room temperature, for which $\pi A < kT$, especially at high pressures and low areas[55].

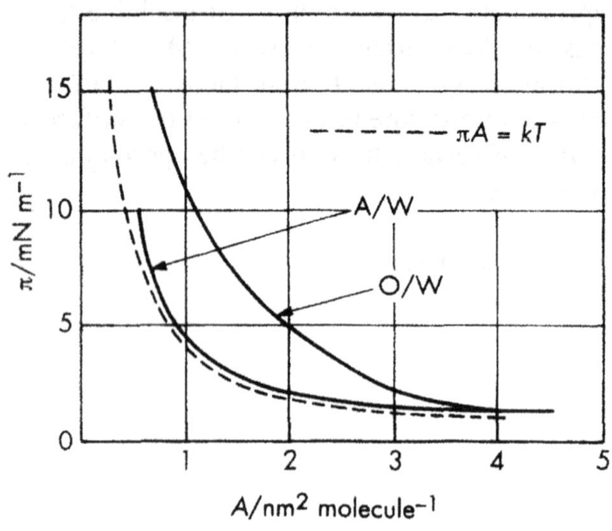

Figure 4.22 π–A curves for cetyl trimethyl ammonium bromide at air–water and oil–water interfaces at 20°C

Condensed films

Palmitic, stearic and higher straight-chain fatty acids are examples of materials which give condensed films at room temperature. At high film areas the molecules of fatty acid do not separate completely from one another, as the cohesion between the hydrocarbon chains is strong enough to maintain the film molecules in small clusters or islands on the surface (Figure 4.23). Because of this strong cohering tendency the surface pressure remains very low as the film is compressed and then rises rapidly when the molecules become tightly packed together.

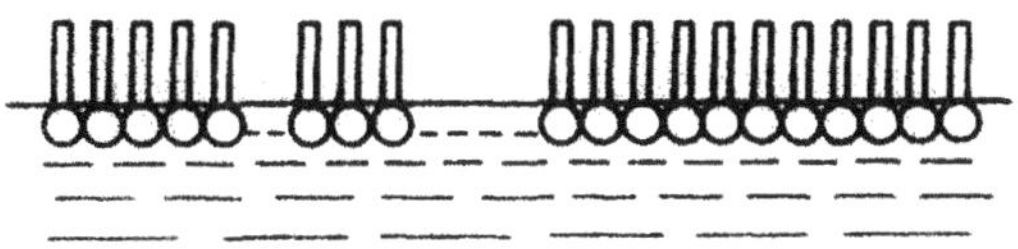

Figure 4.23

For stearic acid spread on dilute HCl an initial pressure rise is observed at about 0.25 nm^2 $molecule^{-1}$, corresponding to the initial packing of the end-groups (Figure 4.24). The π–A curve becomes very steep at about 0.205 nm^2 $molecule^{-1}$, when it is supposed that more efficient packing has been achieved by staggering of the end-groups and interlocking of the hydrocarbon chains. A limiting area of 0.20–0.22 nm^2 $molecule^{-1}$ is observed for straight-chain fatty acids irrespective of the chain length. The packing of the molecules in the film at this point is not far short of that in the crystalline state. The cross-sectional area of stearic acid molecules from X-ray diffraction measurements is about 0.185 nm^2 at normal temperatures. Any attempt to compress a condensed film beyond its limiting area will eventually lead to a collapse or buckling of the film.

Expanded films

Oleic acid (Figure 4.25) gives a much more expanded film than the corresponding saturated acid, stearic acid – i.e. π is greater for any value of A. Because of the double bond there is less cohesion between the hydrocarbon chains than for stearic acid and a greater

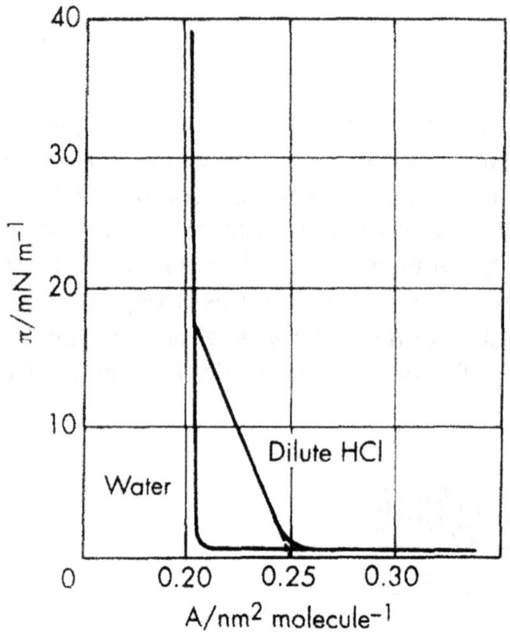

Figure 4.24 π–A curves for stearic acid spread on water and dilute acid substrates at 20°C

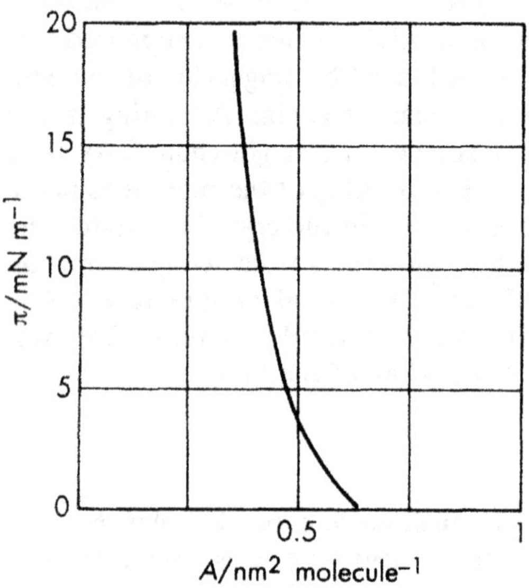

Figure 4.25 π–A curve for oleic acid on water at 20°C

affinity for the aqueous surface. Compression of the oleic acid film forces the double bonds above the surface and eventually orientates the hydrocarbon chains in a vertical position. This process is somewhat gradual, as indicated by the form of the π–A curve. In conformity with this, the rate of oxidation of an oleic acid monolayer by a dilute acid permanganate substrate is found to be greater at high areas.

There are a number of instances in which (with the aid of sensitive measurements) well-defined transitions between gaseous and coherent states are observed as the film is compressed. The π–A curves show a marked resemblance to Andrews' p–V curves for the three-dimensional condensation of vapours to liquids. The π–A curve for myristic acid, given as an example, has been drawn schematically to accentuate its main features (Figure 4.26). Above 8 nm^2 molecule^{-1} the film is gaseous and a liquid-expanded film is obtained on compression to 0.5 nm^2 molecule^{-1}. Fluctuating surface film potentials verify the heterogeneous, transitional nature of the surface between 0.5 nm^2 molecule^{-1} and 8 nm^2 molecule^{-1}.

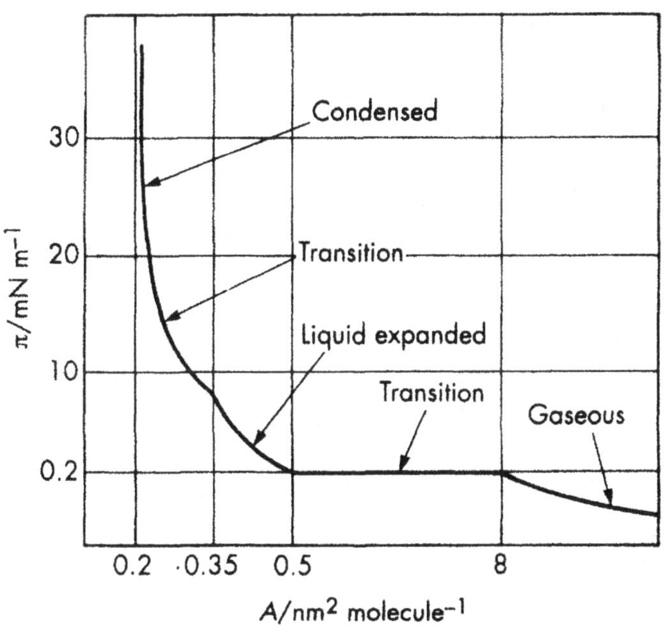

Figure 4.26 Schematic representation of the π–A curve for myristic acid spread on 0.1 mol dm^{-3} HCl at 14°C

Liquid-expanded films obey the equation of state,

$$(\pi - \pi_0)(A - A_0) = kT \tag{4.36}$$

which resembles the van der Waals equation. The accepted theory of the liquid-expanded state, suggested by Langmuir, is that the monolayer behaves as a duplex film in which the head groups are in a state of two-dimensional kinetic agitation while the attractive forces between the hydrocarbon chains keep the film coherent.

Factors influencing the physical state of monomolecular films

As remarked previously, the physical state of a monolayer depends on the lateral cohesive forces between the constituent molecules. By a suitable choice of chain length and temperature, straight-chain fatty acids and alcohols, etc., can be made to exhibit the various monolayer states, one CH_2 group being equivalent to a temperature change of *c*. 5–8 K.

Lateral cohesion also depends on the geometry and orientation of the film molecules, so that the following factors will favour the formation of an expanded film:

1. Bulky head groups, which prevent efficient packing and, hence, maximum cohesion between the hydrocarbon chains.
2. More than one polar group – e.g. unsaturated fatty acids, hydroxy acids. A film pressure is required to overcome the attraction between the second polar group and the aqueous substrate before the molecules can be orientated vertically.
3. More than one hydrocarbon chain orientated in different directions from the polar part of the molecule – e.g. esters, glycerides.
4. Bent hydrocarbon chains – e.g. brassidic acid (*trans*-$CH_3(CH_2)_7CH{=}CH(CH_2)_{11}COOH$), which has a straight hydrocarbon chain, gives a condensed film, whereas erucic acid (*cis*-$CH_3(CH_2)_7CH{=}CH(CH_2)_{11}COOH$), which has a bent hydrocarbon chain, gives a very expanded flim.
5. Branched hydrocarbon chains.

The nature of the substrate, particularly pH, is important when the monolayer is ionisable. When spread on alkaline substrates, because of the ionisation and consequent repulsion between the carboxyl groups, fatty acid monolayers form gaseous or liquid-expanded films at much lower temperatures. Dissolved electrolytes in the substrate can also have a profound effect on the state of the film; for example, Ca^{2+} ions form insoluble calcium soaps with fatty acid films (unless the pH is very low), thus making the film more condensed.

Evaporation through monolayers

Water conservation

The annual loss of water from hot country lakes and reservoirs due to evaporation is usually about 3 m per year. This evaporation can be reduced considerably by coating the water surface with an insoluble monolayer; for example, a monolayer of cetyl alcohol can reduce the rate of evaporation by as much as 40 per cent. Insoluble monolayers also have the effect of damping out surface ripples.

To attain minimum permeability to evaporation, a close-packed monolayer under a sufficient state of compression to squeeze out any surface impurities is required. The monolayer must also be self-healing in response to adverse meteorological conditions such as wind, dust and rainfall, so that spreading ability is also needed. To compromise between these requirements, commercial cetyl alcohol (which also contains some steryl, myristyl and oleyl alcohol) has been used successfully.

The monolayer can be spread initially either from solvent or as a powder, the latter being preferred. Small rafts which allow monolayer material to seep out slowly and replace losses are usually installed at points on the water surface. Oxygen diffuses readily through insoluble monolayers. The oxygen content of the underlying water is somewhat less (80 per cent saturated rather than the normal 90 per cent saturated), since the surface is more quiescent but this has no adverse effect on life beneath the surface.

Monolayers on droplets

Evaporation from small water droplets used to bind fine dust in coal mines is severe unless the surfaces of the droplets are covered with a protective insoluble film. By previously dispersing a little cetyl alcohol in the water the life of the droplets can be increased some thousand-fold. Tarry material, dust, etc., in town fogs is responsible for delay in dispersal because of monolayer formation.

Surface films of proteins

Surface films of high-polymer material, particularly proteins, offer another wide field of study.

Proteins consist of a primary structure of amino acid residues connected in a definite sequence by peptide linkages to form polypeptide chains:

$$-\overset{\text{R}}{\overset{|}{\text{C}}}\text{H}-\text{NH}-\text{CO}-\overset{\text{R}'}{\overset{|}{\text{C}}}\text{H}-\text{NH}-\text{CO}-\overset{\text{R}''}{\overset{|}{\text{C}}}\text{H}-\text{etc.}$$

which may embrace hundreds of such residues. These polypeptide chains normally take up a helical configuration, which is stabilised mainly by hydrogen bonding between spatially adjacent $-NH-$ and $-CO-$ groups. The helical polypeptide chains of the globular proteins are in turn folded up to give compact and often nearly spherical molecules. This configuration is maintained by hydrogen bonding, van der Waals forces between the non-polar parts, disulphide cross-linking, etc.

Any significant alteration in this arrangement of the polypeptide chains without damage to the primary structure is termed *denaturation*. The common agents of denaturation are those which would be expected to modify hydrogen bonds and other weak stabilising linkages – e.g. acids, alkalis, alcohol, urea, heat, ultraviolet light and surface forces. Protein denaturation is accompanied by a marked loss of solubility and is usually, but not necessarily, an irreversible process. Proteins adsorb and denature at high-energy air–water and oil–water interfaces because unfolding allows the polypeptide chains to be orientated with most of their hydrophilic groups in the water phase and most of the hydrophobic groups away from the water phase.

If a small amount of protein solution is suitably spread at the surface of an aqueous substrate, most of the protein will be surface-denatured, giving an insoluble monomolecular film before it has a chance to dissolve. The techniques already described for studying spread monolayers of insoluble material can, therefore, be used in

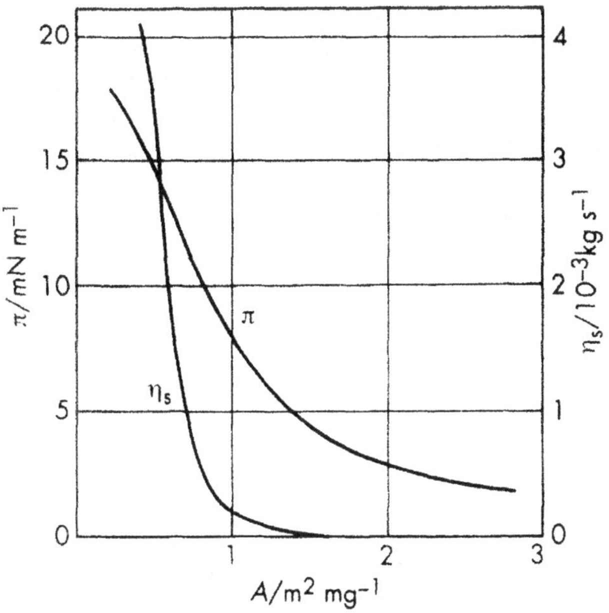

Figure 4.27 π–A and η_s–A curves for a spread monolayer of β-globulin at the petrol ether–water interface[159] (*By courtesy of The Faraday Society*)

the study of protein films. One frequently used spreading solution contains about 0.1 per cent protein in a mixture of alcohol and aqueous sodium acetate.

Compression of protein films below about 1 $m^2\ mg^{-1}$ results in close packing of the polypeptide chains and the gradual development of a gel-like structure. At a surface pressure of *c.* 15–20 $mN\ m^{-1}$ a time-dependent collapse of the film into bundles of insoluble fibres takes place. The area occupied by the compressed protein film is often characterised by the limiting area obtained by extrapolating the approximately linear part of the π–A curve to zero pressure. A more satisfactory characterisation of the close-packed film, especially in relation to the areas indicated by X-ray diffraction measurements on protein fibres, is given by the area at minimum compressibility – i.e. where the π–A curve is at its steepest. This area corresponds approximately to the onset of film collapse. Not surprisingly, these areas show little variation from protein to protein, since the average size of the constituent amino acid residues does not alter greatly.

Many of the proteins from which these films can be formed are approximately spherical in the native state, with diameters of *c.* 5–10 nm. Since a limiting area of 1 m^2 mg^{-1} corresponds to about 0.15 nm^2 per peptide residue, or a film only 0.8–1.0 nm thick, then clearly some unfolding of the polypeptide chains takes place at the surface. Proteins unfold even further at oil–water interfaces.

At low compressions, up to *c.* 1 mN m^{-1}, protein films tend to be gaseous, thus permitting relative molecular mass determination.

For an ideal gaseous film,

$$\pi A_m = RT$$

where A_m is the molar area of the film material. Therefore,

$$\pi AM = RT$$

where A represents the area per unit mass and M is the molar mass of the film material. To realise ideal gaseous behaviour experimental data must be extrapolated to zero surface pressure – i.e.

$$\lim_{\pi \to 0} \pi A = \frac{RT}{M} \tag{4.37}$$

The relative molecular masses of a number of spread proteins have been determined from surface-pressure measurements[54]. In many cases they compare with the relative molecular masses in bulk solution. Relative molecular masses significantly lower than the bulk-solution values have also been reported, which suggests surface dissociation of the protein molecules into sub-units.

Interactions in mixed films

Mixed surface films, especially those likely to be of biological importance, have been subject to a great deal of investigation.

Often there is evidence of interaction between stoichiometric proportions of the components of a mixed monolayer. Evidence of interaction can be sought by measuring partial molecular areas or by studying the collapse of the mixed film. The partial molecular areas of the components of a mixed film are usually different from the molecular areas of the pure components when interaction occurs. A mixed monolayer may collapse in one of two ways: (*a*) with no interaction one component displaces the other, usually at the collapse pressure of the displaced material; or (*b*) interacting mixed films collapse as a whole at a surface pressure different from, and usually greater than, the collapse pressure of either component.

Another type of interaction is the penetration of a surface-active constituent of the substrate into a spread monolayer. Penetration effects can be studied by injecting a solution of the surface-active material into the substrate immediately beneath the monolayer: (*a*) if there is no association between the injected material and the monolayer, π and ΔV will both remain unaltered; (*b*) if the injected material adsorbs on to the underside of the monolayer without actual penetration, ΔV will change appreciably but π will alter very little; (*c*) if the injected material penetrates into the monolayer (i.e. when there is association between both polar and non-polar parts of the injected and original monolayer materials), π will change significantly and ΔV will assume an intermediate value between ΔV of the original monolayer and ΔV of a monolayer of injected material. Penetration is less likely to occur when the monolayer is tightly packed.

It is often desirable (e.g. in emulsions) to have a surface film of charged species; however, as such, repulsion between the head groups usually makes such a film non-coherent. Surface films which are both charged and coherent can be produced by use of a mixture of ionic and non-ionic surfactants, especially where the structural characteristics of the surfactant molecules are such that they pack efficiently between one another.

Biological membranes[59–61]

Biological membranes consist mainly of lipoprotein material. The classic experiment relating to membrane structure was first performed by Gorter and Grendel[160]. They extracted the lipid from erythrocyte membranes, spread it at an air–water interface and found that the film compressed to a limiting area which corresponded to twice the external area of the cells from which the lipid was derived. The results of experiments such as this led Danielli and Davson[161] to propose that the cell membrane consists essentially of a bimolecular layer of lipid with the hydrocarbon chains orientated towards the interior and the hydrophilic groups on the outside. As with micelles (see pages 85 and 88), the organisation of such a cell membrane is primarily the result of hydrophobic bonding[49]. The permeability of the cell membrane to specific metabolites, however, requires modification of the simple lipid bimolecular layer model, and the basic structure is thought to consist of the lipid bimolecular layer with protein both adsorbed on to and incorporated into it[62], as illustrated in Figure 4.28.

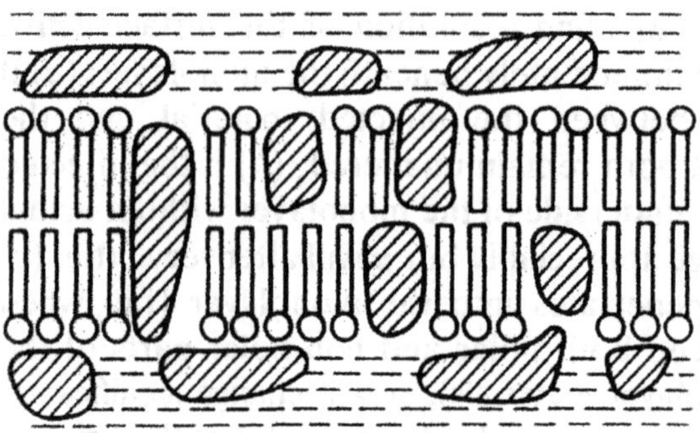

Figure 4.28 Schematic representation of a biological cell membrane. A bimolecular layer of phospholipid with hydrocarbon chains orientated to the interior and hydrophilic groups on the outside is penetrated by protein (shaded areas). Protein is also found adsorbed at the membrane surface

Artificially prepared monolayers have been used successfully as models for studying certain biological processes. Schulman and Rideal[162] investigated the action of agglutinating (coagulating) and lytic agents on red blood cells by the monolayer technique. A mixed film of 20 per cent cholesterol and 80 per cent gliadin was used to represent the red blood cell membrane. If was found that lytic agents penetrate into this monolayer, whereas agglutinating agents only adsorb on to the underside of the film. This implies that red blood cells are altered structurally (lysis) owing to the penetration of lytic agents into the membrane, while agglutinating agents merely adsorb on to the membrane surface.

5 The solid–gas interface

Adsorption of gases and vapours on solids

When a gas or vapour is brought into contact with a clean solid surface, some of it will become attached to the surface in the form of an adsorbed layer. The solid is generally referred to as the *adsorbent*, adsorbed gas or vapour as the *adsorbate* and non-adsorbed gas as the *adsorptive*. It is possible that uniform absorption into the bulk of the solid might also take place, and, since *adsorption* and *absorption* cannot always be distinguished experimentally, the generic term *sorption* is sometimes used to describe the general phenomenon of gas uptake by solids.

Any solid is capable of adsorbing a certain amount of gas, the extent of adsorption at equilibrium depending on temperature, the pressure of the gas and the effective surface area of the solid. The most notable adsorbents are, therefore, highly porous solids, such as charcoal and silica gel (which have large internal surface areas – up to *c*. 1000 $m^2\ g^{-1}$) and finely divided powders. The relationship at a given temperature between the equilibrium amount of gas adsorbed and the pressure of the gas is known as the adsorption isotherm (Figures 5.1, 5.5–5.6, 5.8, 5.11, 5.13).

Adsorption reduces the imbalance of attractive forces which exists at a surface, and, hence, the surface free energy of a heterogeneous system. In this respect, the energy considerations relating to solid surfaces are, in principle, the same as those already discussed for liquid surfaces. The main differences between solid and liquid surfaces arise from the fact that solid surfaces are heterogeneous in respect of activity, with properties dependent, to some extent, on previous environment.

Physical adsorption[2,3,63–4] and chemisorption[2,65–68]

The forces involved in the adsorption of gases and vapours by solids may be non-specific (van der Waals) forces, similar to the forces involved in liquefaction, or stronger specific forces, such as those which are operative in the formation of chemical bonds. The former are responsible for *physical adsorption* and the latter for *chemisorption*.

When adsorption takes place, the gas molecules are restricted to two-dimensional motion. Gas adsorption processes are, therefore, accompanied by a decrease in entropy. Since adsorption also involves a decrease in free energy, then, from the thermodynamic relationship,

$$\Delta G = \Delta H - T\Delta S \tag{5.1}$$

it is evident that $\Delta H_{\text{ads.}}$ must be negative – i.e. the adsorption of gases and vapours on solids is always an exothermic process.* The extent of gas adsorption (under equilibrium conditions), therefore, increases with decreasing temperature (see Figure 5.1). Heats of adsorption can be measured by direct calorimetric methods. Isosteric (constant adsorption) heats of adsorption can be calculated from reversible adsorption isotherms using the Clausius–Clapeyron equation:

$$\left(\frac{\partial \ln p}{\partial T}\right)_V = \frac{-\Delta H_{\text{ads.}}}{RT^2} \tag{5.2}$$

The heats of physical adsorption of gases are usually similar to their heats of condensation; for example, the integral heat of physical adsorption of nitrogen on an iron surface is *c.* -10 kJ mol^{-1} (the heat of liquefaction of nitrogen is -5.7 kJ mol^{-1}). Heats of chemisorption are, in general, much larger; for example, the integral heat of chemisorption of nitrogen on iron is *c.* -150 kJ mol^{-1}.

The attainment of physical adsorption equilibrium is usually rapid, since there is no activation energy involved, and (apart from complications introduced by capillary condensation) the process is readily reversible. Multilayer physical adsorption is possible, and at

*This argument does not necessarily hold for processes such as adsorption from solution and micellisation, since a certain amount of destructuring (e.g. desolvation) may be involved and the net entropy change may be positive.

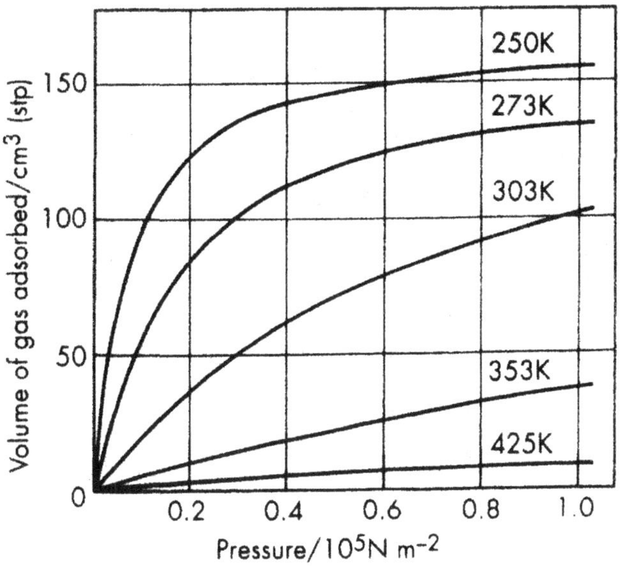

Figure 5.1 Adsorption isotherms for ammonia on charcoal[163]

the saturated vapour pressure of the gas in question physical adsorption becomes continuous with liquefaction.

Only monomolecular chemisorbed layers are possible. Chemisorption is a specific process which may require an activation energy and may, therefore, be relatively slow and not readily reversible. The nature of physical adsorption and chemisorption is illustrated by the schematic potential energy curves shown in Figure 5.2 for the adsorption of a diatomic gas X_2 on a metal M.

Curve P represents the physical interaction energy between M and X_2. It inevitably includes a short-range negative (attractive) contribution arising from London–van der Waals dispersion forces and an even shorter-range positive contribution (Born repulsion) due to an overlapping of electron clouds. It will also include a further van der Waals attractive contribution if permanent dipoles are involved. The nature of van der Waals forces is discussed on page 215.

Curve C represents chemisorption, in which the adsorbate X_2 dissociates to $2X$. For this reason, an energy equal to the dissociation energy of X_2 is represented at large distances. The curve is also characterised by a relatively deep minimum which represents the heat of chemisorption, and which is at a shorter distance from the solid

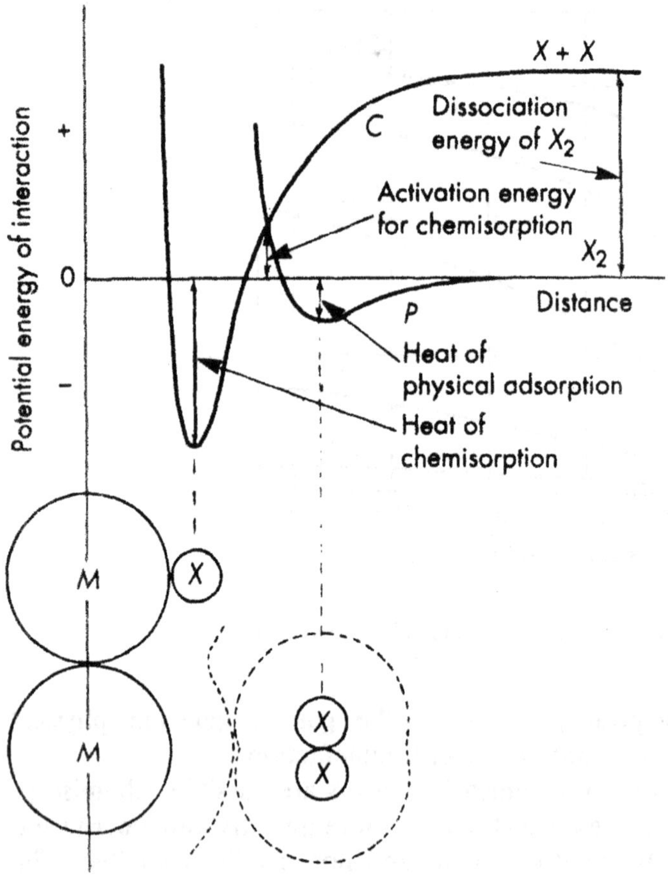

Figure 5.2 Potential energy curves for physical adsorption and chemisorption

surface than the relatively shallow minimum in the physical adsorption curve.

It can be seen from these curves that initial physical adsorption is a most important feature of chemisorption. If physical adsorption were non-existent, the energy of activation for chemisorption would be equal to the high dissociation energy of the adsorbate gas molecules. As it is, an adsorbate gas molecule is first physically adsorbed, which involves approaching the solid surface along a low-energy path. Transition from physical adsorption to chemisorption takes place at the point where curves *P* and *C* intersect, and the energy at this point is equal to the activation energy for chemisorption. The magnitude of this activation energy depends, therefore, on the shapes of the physical adsorption and chemisorption curves, and varies widely from system to system; for example, it is low for the chemisorption of hydrogen on to most metal surfaces.

If the activation energy for chemisorption is appreciable, the rate of chemisorption at low temperature may be so slow that, in practice, only physical adsorption is observed.

Figure 5.3 shows how the extent of gas adsorption on to a solid surface might vary with temperature at a given pressure. Curve (a) represents physical adsorption equilibrium and curve (b) represents chemisorption equilibrium. The extent of adsorption at temperatures at which the rate of chemisorption is slow, but not negligible, is represented by a non-equilibrium curve, such as (c), the location of which depends on the time allowed for equilibrium.

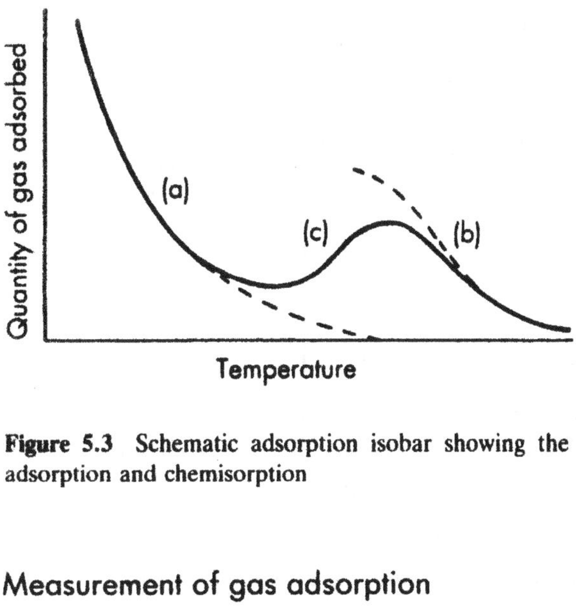

Figure 5.3 Schematic adsorption isobar showing the transition between physical adsorption and chemisorption

Measurement of gas adsorption

Usually, the solid adsorbent under investigation must first be freed, as far as possible from previously adsorbed gases and vapours. Evacuation to *c*. 10^{-4} Torr (outgassing) for several hours will usually remove physically adsorbed gas. It is difficult, and often impossible, to remove chemisorbed gas completely unless the solid is heated to a high temperature (*c*. 100–400°C). Such treatment might cause sintering and an alteration of the sorptive capacity of the solid.

The adsorption of a gas or vapour can be measured by admitting a known amount of the adsorbate into an evacuated, leak-free space containing the outgassed adsorbent. The extent of adsorption can then be determined either volumetrically or gravimetrically.

The volumetric method is mainly used for the purpose of determining specific surface areas of solids from gas (particularly nitrogen) adsorption measurements (see page 134). The gas is contained in a gas burette, and its pressure is measured with a manometer (see Figure 5.4). All of the volumes in the apparatus are calibrated so that when the gas is admitted to the adsorbent sample the amount adsorbed can be calculated from the equilibrium pressure reading. The adsorption isotherm is obtained from a series of measurements at different pressures.

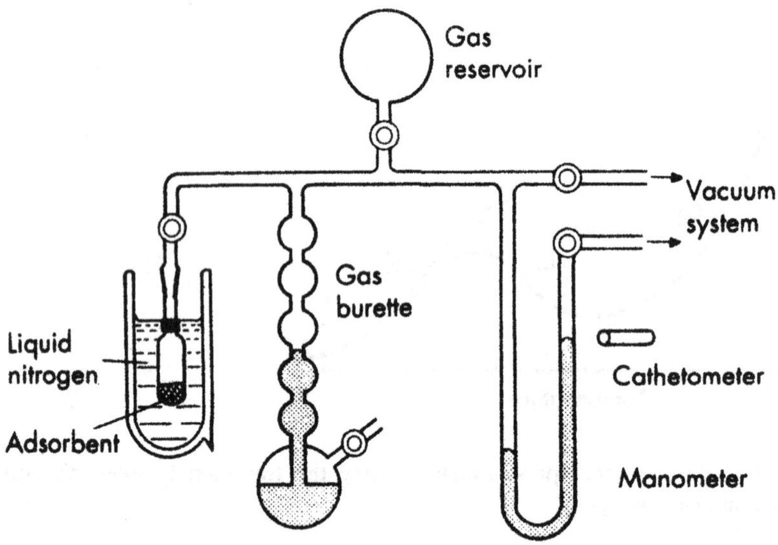

Figure 5.4 Volumetric apparatus for measuring gas adsorption at 77 K

The gravimetric method can be used for studying both gas and vapour sorption. Purified gas or vapour is introduced into an evacuated apparatus in which the sorbent sample is contained in a weighing pan, the pressure is noted and the extent of sorption is measured directly (allowing for buoyancy) as the increase in the weight of the sorbent sample. In the original method of McBain and Bakr[164] a precalibrated quartz spiral is used for weighing the sample, but this technique has largely been superseded by the use of beam-type vacuum microbalances.

An alternative gas-adsorption method, which does not require vacuum equipment, is that developed by Nelsen and Eggersten[165]. It bears some similarity to gas chromatography. A gas mixture containing the adsorbate (usually nitrogen) and a carrier gas (usually helium) is passed over the solid under test at room temperature. The gas flow into and out of the sample container is monitored by means of a pair of thermal conductivity detectors. When equilibrium has been established, the sample is cooled (e.g. by immersion in liquid nitrogen). Owing to gas adsorption, the outlet stream is depleted for a time in adsorbate, the thermal conductivity detectors are thrown off balance and the amount of gas adsorbed can be measured in terms of the area under a peak on a recording potentiometer. On warming the sample, desorption takes place and a negative peak of equal area is given. This technique is extremely useful for surface area determination by the BET method (see page 134).

Classification of adsorption isotherms

Three phenomena may be involved in physical adsorption:

1. Monomolecular adsorption.
2. Multimolecular adsorption.
3. Condensation in pores or capillaries.

Frequently, there is overlapping of these phenomena, and the interpretation of adsorption studies can be complicated. Brunauer has classified adsorption isotherms into the five characteristic types shown in Figure 5.5. To facilitate comparison of isotherms, it is preferable to plot them in terms of relative pressures (p/p_0), where p_0 is the saturation vapour pressure, rather than pressure itself. This also has the advantage of giving a 0 to 1 scale for all gases.

Type I isotherms (e.g. ammonia on charcoal at 273 K) show a fairly rapid rise in the amount of adsorption with increasing pressure up to a limiting value. They are referred to as Langmuir-type isotherms and are obtained when adsorption is restricted to a monolayer. Chemisorption isotherms, therefore, approximate to this shape. Type I isotherms have also been found for physical adsorption on solids

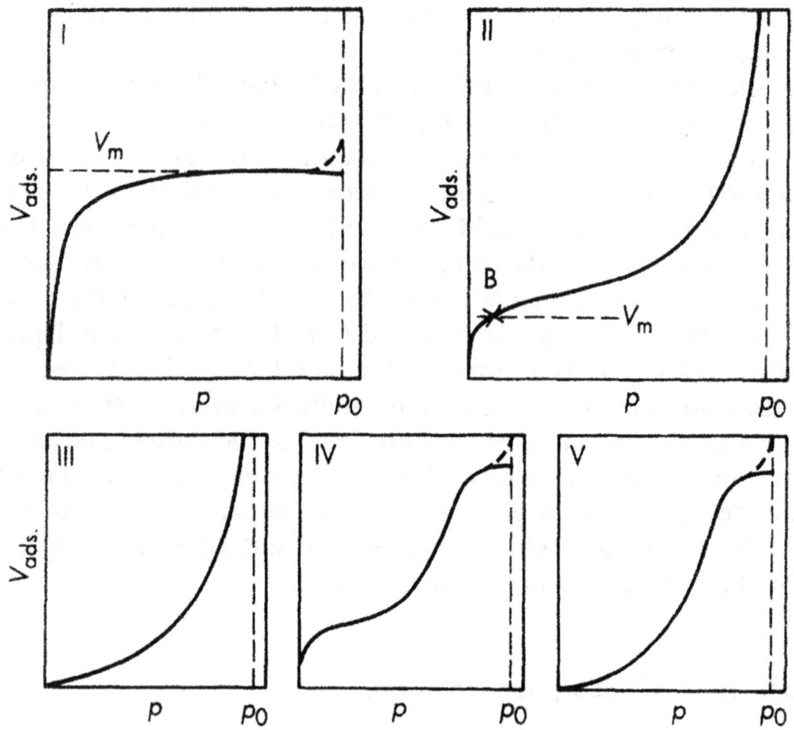

Figure 5.5 Brunauer's classification of adsorption isotherms (p_0 = saturated vapour pressure)

containing a very fine pore structure (e.g. nitrogen on microporous carbon at 77 K).

Type II isotherms (e.g. nitrogen on silica gel at 77 K) are frequently encountered, and represent multilayer physical adsorption on non-porous solids. They are often referred to as sigmoid isotherms. For such solids, point B represents the formation of an adsorbed monolayer. Physical adsorption on microporous solids can also result in type II isotherms. In this case, point B represents both monolayer adsorption on the surface as a whole and condensation in the fine pores. The remainder of the curve represents multilayer adsorption as for non-porous solids.

Type IV isotherms (e.g. benzene on iron(III) oxide gel at 320 K) level off near the saturation vapour pressure and are considered to reflect capillary condensation in porous solids, the effective pore diameters usually being between 2 nm and 20 nm. The upper limit of adsorption is mainly governed by the total pore volume.

Types III (e.g. bromine on silica gel at 352 K) and V (e.g. water vapour on charcoal at 373 K) show no rapid initial uptake of gas, and occur when the forces of adsorption in the first monolayer are relatively small. These isotherms are rare.

Many adsorption isotherms are borderline cases between two or more of the above types. In addition, there are some isotherms which do not fit into Brunauer's classification at all, the most notable being the stepwise isotherms, an example of which is given in Figure 5.6. Stepwise isotherms are usually associated with adsorption on to uniform solid surfaces, each step corresponding to the formation of a complete monomolecular adsorbed layer (see page 133).

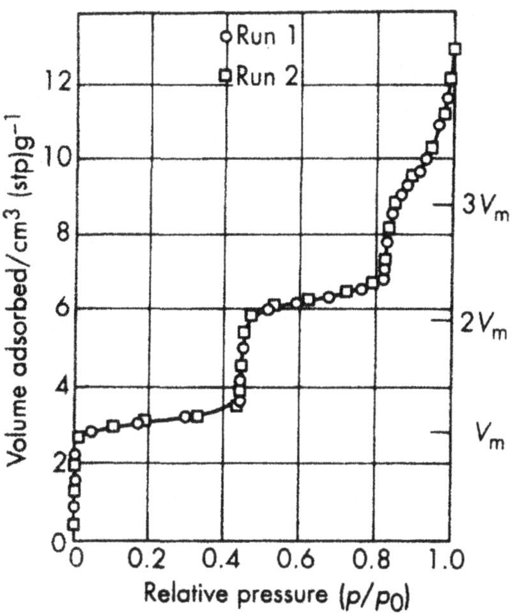

Figure 5.6 Stepwise isotherm for the adsorption of krypton at 90 K on carbon black (graphitised at 3000 K)[166] (*By courtesy of the Canadian Journal of Chemistry*)

Capillary condensation

Most solids are porous to some extent. Porosity may be the result of gas evolution during the formation (geological or industrial) of the solid, it may reflect a fibrous structure, or it may be the result of compaction of particulate solid.

A most interesting, and useful, class of porous solids is the zeolites[69]. These are materials (natural and synthetic) in which SiO_4 and AlO_4 tetrahedra are linked by sharing oxygen atoms to give ring structures, which, in turn, are linked to give an overall three-dimensional structure which contains regular channels and cavities of sizes similar to those of small to medium-sized molecules. As such, they are ideal molecular sieves, and, in view of the AlO_4^- acid sites they contain, a high level of selectivity can be achieved in catalysis and ion-exchange.

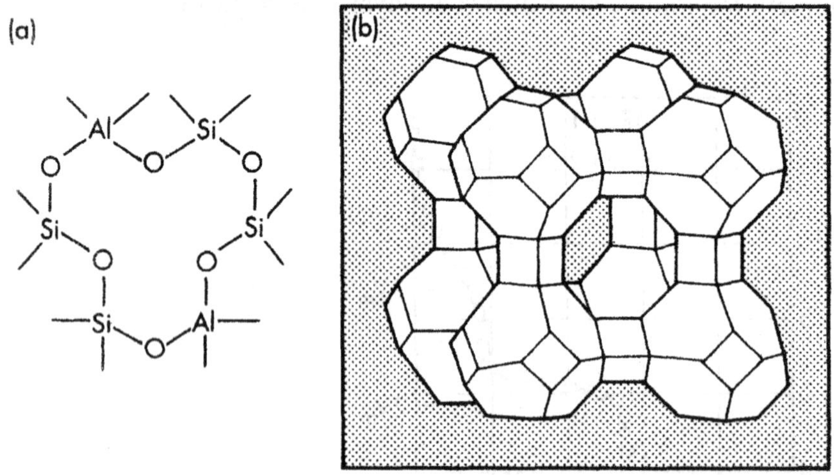

Figure 5.7 Zeolite structure. (a) 6-ring containing two aluminium and four silicon tetrahedral centres. (b) Zeolite A structure. Each of the eight sodalite units depicted contains 24 aluminium or silicon tetrahedral centres arranged to give six 4-rings plus eight 6-rings

Pores are usually classified according to their widths as follows:

Micropores	< 2 nm
Mesopores	2 nm to 50 nm
Macropores	> 50 nm

It has previously been shown (page 68) that the vapour pressure over a convex liquid surface is greater than that over the corresponding flat surface. A liquid which wets the wall of a capillary will have a concave liquid-vapour interface and, therefore, a lower vapour pressure in the capillary than it would have over a flat surface. This vapour pressure difference is given by the Kelvin equation, written in the form

$$RT \ln p_r / p_0 = -\frac{2\gamma V_m \cos\theta}{r} \tag{5.3}$$

where r is the radius of the capillary, and θ the contact angle between the liquid and the capillary wall.

Condensation can, therefore, take place in narrow capillaries at pressures which are lower than the normal saturation vapour pressure. Zsigmondy (1911) suggested that this phenomenon might also apply to porous solids. Capillary rise in the pores of a solid will usually be so large that the pores will tend to be either completely full of capillary condensed liquid or completely empty. Ideally, at a certain pressure below the normal condensation pressure all the pores of a certain size and below will be filled with liquid and the rest will be empty. It is probably more realistic to assume that an adsorbed monomolecular film exists on the pore walls before capillary condensation takes place. By a corresponding modification of the pore diameter, an estimate of pore size distribution (which will only be of statistical significance because of the complex shape of the pores) can be obtained from the adsorption isotherm.

Capillary condensation is also important in the binding of dust and powder particles by water. Particles separated by a thin layer of water are held together very strongly by capillary forces. The inhibition of evaporation due to the concave shape of the air–water interface enhances the duration of this particle binding.

The capillary condensation theory provides a satisfactory explanation of the phenomenon of adsorption hysteresis, which is frequently observed for porous solids. 'Adsorption hysteresis' is a term which is used when the desorption isotherm curve does not coincide with the adsorption isotherm curve (Figure 5.8).

A possible explanation of this phenomenon is given in terms of contact angle hysteresis. The contact angle on adsorption, when liquid is advancing over a dry surface, is usually greater than the contact angle during desorption, when liquid is receding from a wet surface. From the Kelvin equation, it is evident that the pressure below which liquid vaporises from a particular capillary will, under

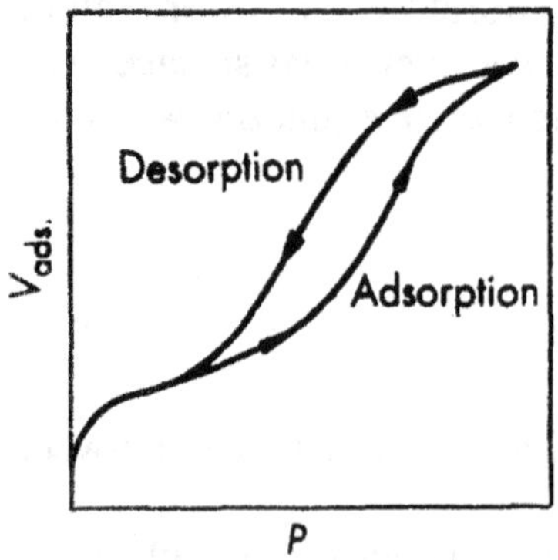

Figure 5.8 A hysteresis loop in physical adsorption

these circumstances, be lower than the pressure required for capillary condensation.

Another theory of adsorption hysteresis considers that there are two types of pores present, each having a size distribution. The first type are V-shaped, and these fill and empty reversibly. The second type have a narrow neck and a relatively wide interior. These 'ink-bottle' pores are supposed to fill completely when a p/p_0 value corresponding to the relatively wide pore interior is reached, but once filled they retain their contents until p/p_0 is reduced to a value corresponding to the relatively small width of the pore neck.

In a further theory, the pores are considered to be open-ended cylinders. Condensation will commence on the pore walls, for which the principal radii of curvature are the pore radius and infinity, and continue until the pore is filled with condensed liquid. Evaporation must take place from the concave liquid surfaces at the ends of the pore, for which (assuming zero contact angle) the principal radii of curvature are both equal to the pore radius.

Mercury intrusion porosimetry[70]

Pore size distributions are often determined by the technique of mercury intrusion porosimetry. The volume of mercury (contact angle *c*. 140° with most solids) which can be forced into the pores of the solid is measured as a function of pressure. The pore size distribution is calculated in accordance with the equation for the pressure difference across a curved liquid interface,

$$p = 2\gamma \cos \theta \, / \, r \tag{5.4}$$

If all the pores are equally accessible, only those for which r is greater than $\gamma \cos \theta/p$ will be filled, i.e. each pressure increment causes a group of smaller pores to be filled and a cumulative pore volume as a function of pore size can be determined, as illustrated in Figure 5.9.

When the mercury pressure is reduced, hysteresis is usually observed. This will reflect some of the mercury being permanently trapped in 'ink-bottle' pores and, as such, the 'ink-bottle' pore volume is given by the residual mercury entrapped when the mercury pressure is reduced to atmospheric pressure. Hysteresis, however, can also result from structural damage sustained due to the very high mercury pressures involved.

A related technique for measuring the contact angle of a liquid with a porous or finely divided solid is described on page 157.

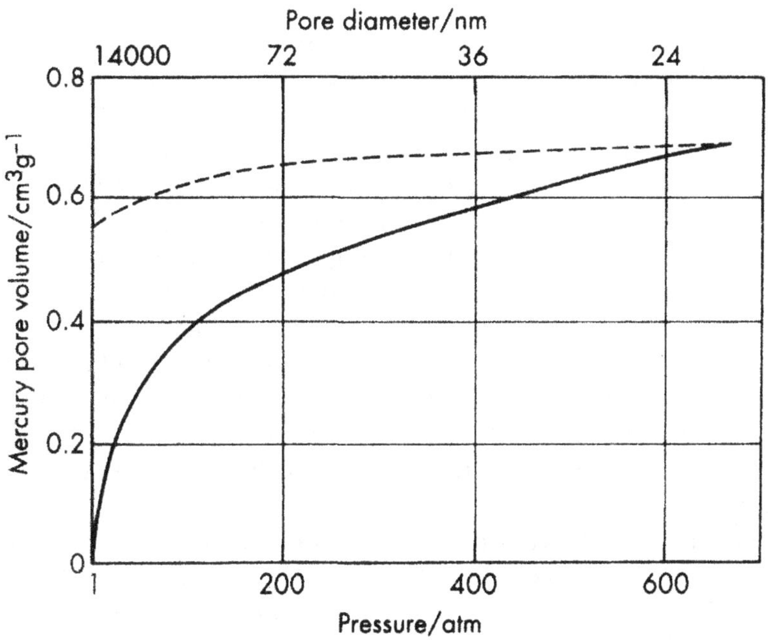

Figure 5.9 Mercury pressurisation (full line) and depressurisation (broken line) for a sample of activated carbon

Isotherm equations

Numerous attempts have been made at developing mathematical expressions from postulated adsorption mechanisms to fit the various experimental isotherm curves. The three isotherm equations which are most frequently used are those due to Langmuir, to Freundlich, and to Brunauer, Emmett and Teller (BET).

The Langmuir adsorption isotherm

Before 1916, adsorption theories postulated either a condensed liquid film or a compressed gaseous layer which decreases in density as the distance from the surface increases. Langmuir (1916) was of the opinion that, because of the rapidity with which intermolecular forces fall off with distance, adsorbed layers are not likely to be more than one molecular layer in thickness. This view is generally accepted for chemisorption and for physical adsorption at low pressures and moderately high temperatures.

The Langmuir adsorption isotherm is based on the characteristic assumptions that (*a*) only monomolecular adsorption takes place, (*b*) adsorption is localised and (*c*) the heat of adsorption is independent of surface coverage. A kinetic derivation follows in which the velocities of adsorption and desorption are equated with each other to give an expression representing adsorption equilibrium.

Let V equal the equilibrium volume of gas adsorbed per unit mass of adsorbent at a pressure p and V_{m} equal the volume of gas required to cover unit mass of adsorbent with a complete monolayer.

The velocity of adsorption depends on: (*a*) the rate at which gas molecules collide with the solid surface, which is proportional to the pressure; (*b*) the probability of striking a vacant site $(1-V/V_{\mathrm{m}})$; and (*c*) an activation term $\exp[-E/RT]$, where E is the activation energy for adsorption.

The velocity of desorption depends on: (*a*) the fraction of the surface which is covered, V/V_{m}; and (*b*) an activation term $\exp[-E'/RT]$, where E' is the activation energy for desorption.

Therefore, when adsorption equilibrium is established,

$$p(1-V/V_{\mathrm{m}})\exp[-E/RT]=k(V/V_{\mathrm{m}})\exp[-E'/RT]$$

where k is a proportionality constant – i.e.

$$p = k\exp[\Delta H_{\text{ads.}} / RT]\frac{V / V_{\text{m}}}{(1 - V / V_{\text{m}})}$$

where $\Delta H_{\text{ads.}} = E - E'$ = heat of adsorption (negative).

Assuming that the heat of adsorption, $\Delta H_{\text{ads.}}$, is independent of surface coverage,

$$k\exp[\Delta H_{\text{ads.}} / RT] = 1 / a$$

where a is a constant dependent on the temperature, but independent of surface coverage. Therefore,

$$ap = \frac{V / V_{\text{m}}}{(1 - V / V_{\text{m}})} \tag{5.5}$$

or $$V = \frac{V_{\text{m}}ap}{(1 + ap)} \tag{5.6}$$

or $$p / V = p / V_{\text{m}} + 1 / aV_{\text{m}} \tag{5.7}$$

i.e. a plot of p/V versus p should give a straight line of slope $1/V_{\text{m}}$ and an intercept of $1/aV_{\text{m}}$ on the p/V axis.

At low pressures the Langmuir isotherm equation reduces to $V = V_{\text{m}}ap$ – i.e. the volume of gas adsorbed varies linearly with pressure. At high pressures a limiting monolayer coverage, $V = V_{\text{m}}$, is reached. The curvature of the isotherm at intermediate pressures depends on the value of the constant a and, hence, on the temperature.

The most notable criticism of the Langmuir adsorption equation concerns the simplifying assumption that the heat of adsorption is independent of surface coverage, which, as discussed in the next section, is not likely to be the case. Nevertheless, many experimental adsorption isotherms fit the Langmuir equation reasonably well.

When the components of a gas mixture compete for the adsorption sites on a solid surface, the Langmuir equation takes the general form

$$\frac{V_i}{V_{\text{m},i}} = \frac{a_i p_i}{1 + \Sigma a_j p_j} \tag{5.8}$$

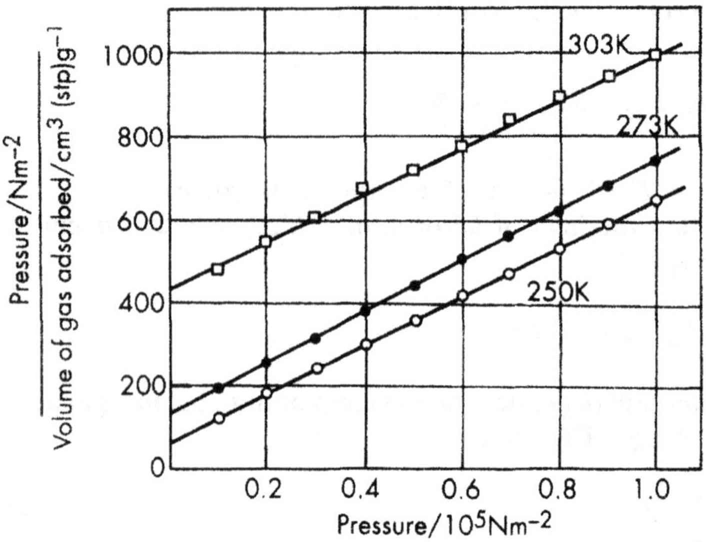

Figure 5.10 Langmuir plots for the adsorption of ammonia on charcoal shown in *Figure 5.1*. Slope = $1/V_m$

The Freundlich (or Classical) adsorption isotherm

The variation of adsorption with pressure can often be represented (especially at moderately low pressures) by the equation

$$V = kp^{1/n} \tag{5.9}$$

where k and n are constants, n usually being greater than unity. Taking logarithms,

$$\log V = \log k + 1/n \log p \tag{5.10}$$

i.e. a plot of log V versus log p should give a straight line.

This adsorption equation was originally proposed on a purely empirical basis. It can be derived theoretically, however, for an adsorption model in which the magnitude of the heat of adsorption varies exponentially with surface coverage. The Freundlich equation is, in effect, the summation of a distribution of Langmuir equations; however, the volume of gas adsorbed is not depicted as approaching a limiting value as in a single Langmuir equation.

The BET equation for multimolecular adsorption

Because the forces acting in physical adsorption are similar to those operating in liquefaction (i.e. van der Waals forces), physical adsorption (even on flat and convex surfaces) is not limited to a monomolecular layer, but can continue until a multimolecular layer of liquid covers the adsorbent surface.

The theory of Brunauer, Emmett and Teller[167] is an extension of the Langmuir treatment to allow for multilayer adsorption on non-porous solid surfaces. The BET equation is derived by balancing the rates of evaporation and condensation for the various adsorbed molecular layers, and is based on the simplifying assumption that a characteristic heat of adsorption ΔH_1 applies to the first monolayer, while the heat of liquefaction, ΔH_L, of the vapour in question applies to adsorption in the second and subsequent molecular layers. The equation is usually written in the form

$$\frac{p}{V(p_0 - p)} = \frac{1}{V_m c} + \frac{(c-1)}{V_m c}\frac{p}{p_0} \tag{5.11}$$

where p_0 is the saturation vapour pressure, V_m is the monolayer capacity and $c \approx \exp[(\Delta H_L - \Delta H_1)/RT]$.

The main purpose of the BET equation is to describe type II isotherms. In addition, it reduces to the Langmuir equation at low pressures; and type III isotherms are given in the unusual circumstances, when monolayer adsorption is less exothermic than liquefaction, i.e. $c < 1$ (see Figure 5.11).

The BET model can also be applied to a situation which might be applicable to porous solids. If adsorption is limited to n molecular layers (where n is related to the pore size), the equation

$$V = \frac{V_m c x}{(1-x)} \cdot \frac{1-(n+1)x^n + nx^{n+1}}{1+(c-1)x - cx^{n+1}} \tag{5.12}$$

is obtained[168], where $x = p/p_0$. This equation is, in fact, a general expression which reduces to the Langmuir equation when $n = 1$ and to the BET equation when $n = \infty$.

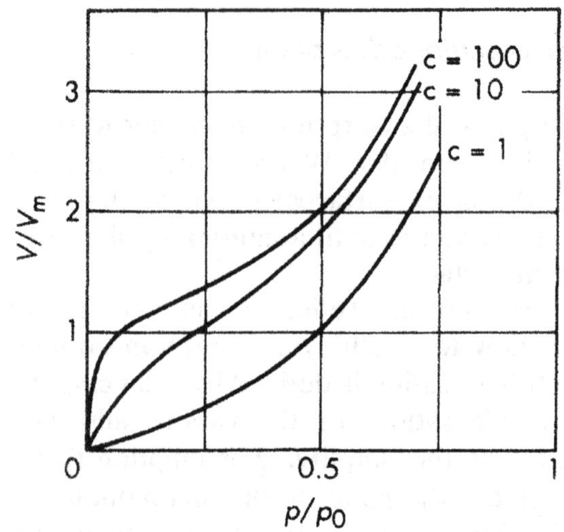

Figure 5.11 BET isotherms

Adsorption energies

A most important feature of the models upon which adsorption isotherm equations, such as those above, are based is a characteristic assumption relating to heat of adsorption and surface coverage. Several factors merit consideration in this respect.

Solid surfaces are usually heterogeneous; therefore, since adsorption at the more active sites is favoured, heats of both monolayer physical adsorption and chemisorption might, in this respect, be expected to become significantly less exothermic as the surface coverage increases, as, for example, shown at low pressures in Figures 5.12a and 5.12b. This, in turn would cause the initial slope of an adsorption isotherm to be steeper than that predicted according to the Langmuir equation or the BET equation.

Chemisorption might involve the adsorbate gas molecules either giving up electrons to or receiving electrons from the adsorbent solid. As either of these processes continues, further adsorption becomes more and more difficult and monolayer coverage is not as readily attained as would be predicted according to the Langmuir equation. The heat of adsorption becomes less exothermic as monolayer coverage is approached, as, for example, shown in Figure 5.12b.

When a gas molecule is adsorbed on to a solid surface which is already partially covered with a monomolecular layer, lateral interaction with the adsorbed gas molecules will be involved in addition to interaction with the solid. In this respect, the heat of adsorption might be expected to become more exothermic with increasing surface coverage, as, for example, shown in Figure 5.12c.

The shape of a multilayer physical adsorption isotherm depends on the tendency of each adsorbed monomolecular layer (particularly the first layer) to be completed before any adsorption into further layers takes place. This situation is favoured if the adsorption energy for the layer being completed is significantly more exothermic than that for commencing further adsorbed layers. As a rather extreme example, Figure 5.12c shows this kind of variation of adsorption energy with surface coverage for the physical adsorption of a gas on a fairly homogeneous solid surface. The corresponding adsorption isotherm

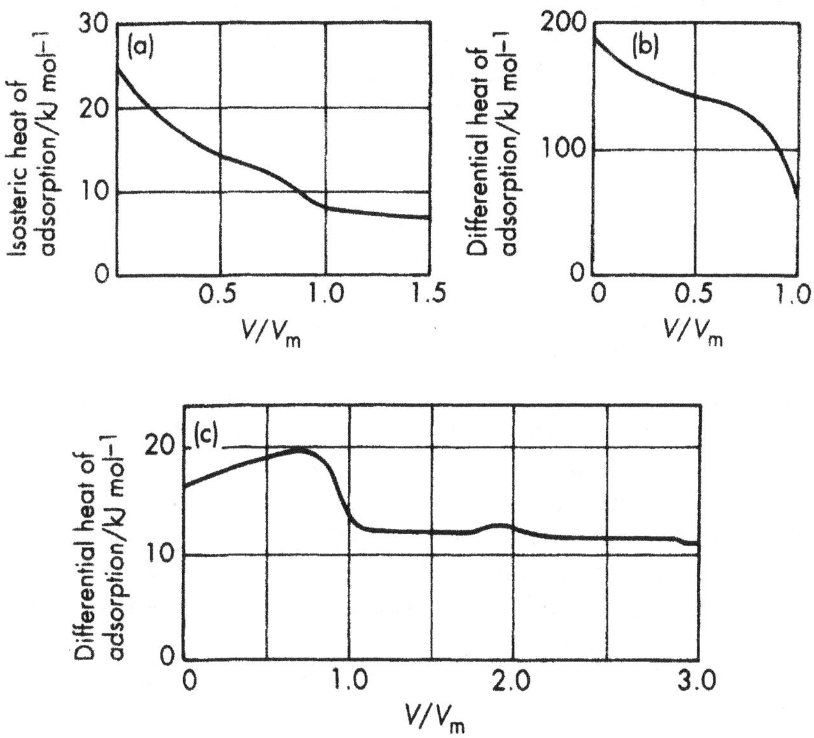

Figure 5.12 Adsorption energy and surface coverage. (a) Physical adsorption of nitrogen on rutile at 85 K[71]. (b) Chemisorption of hydrogen on tungsten[169]. (c) Physical adsorption of krypton on graphitised carbon black[166]. (*See Figure 5.6*) (*By courtesy of (a) Science Progress, (b) Discussions of the Faraday Society and (c) The Canadian Journal of Chemistry*)

(Figure 5.6) shows at least two distinct steps, each corresponding to the formation of an adsorbed monomolecular layer. In most cases of multilayer physical gas adsorption, however, the adsorption energies are such that there is a greater or lesser tendency for the first monomolecular adsorbed layer to be completed prior to any adsorption into the second monolayer, but little tendency for the second monolayer to be completed prior to adsorption into the third and subsequent monolayers.

Surface areas

The monolayer capacity, V_m, is a parameter of particular interest, since it can be used for calculating the surface area of an adsorbent if the effective area occupied by each adsorbate molecule is known.

If the BET equation is applicable to a multilayer physical adsorption isotherm, a plot of $p/V(p_0-p)$ versus p/p_0 gives a straight line of slope $(c-1)/V_mc$ and an intercept of $1/V_mc$ on the $p/V(p_0-p)$ axis – i.e.

$$V_m = \frac{1}{\text{slope} + \text{intercept}} \tag{5.13}$$

Using the appropriate gas equation, the monolayer capacity can be calculated in terms of adsorbed molecules per unit mass of adsorbent.

Although the BET equation is open to a great deal of criticism, because of the simplified adsorption model upon which it is based, it nevertheless fits many experimental multilayer adsorption isotherms particularly well at pressures between about 0.05 p_0 and 0.35 p_0 (within which range the monolayer capacity is usually reached). However, with porous solids (for which adsorption hysteresis is characteristic), or when point B on the isotherm (Figure 5.5) is not very well defined, the validity of values of V_m calculated using the BET equation is doubtful.

With a small loss of accuracy, the straight-line BET plot (Figure 5.13) can be assumed to pass through the origin and V_m can be calculated on the basis of a single gas-adsorption measurement (usually with p/p_0 between 0.2 and 0.3). This procedure is frequently adopted for routine surface area measurements.

The adsorbate most commonly used for BET surface area determinations is nitrogen at 77 K (liquid nitrogen temperature). The

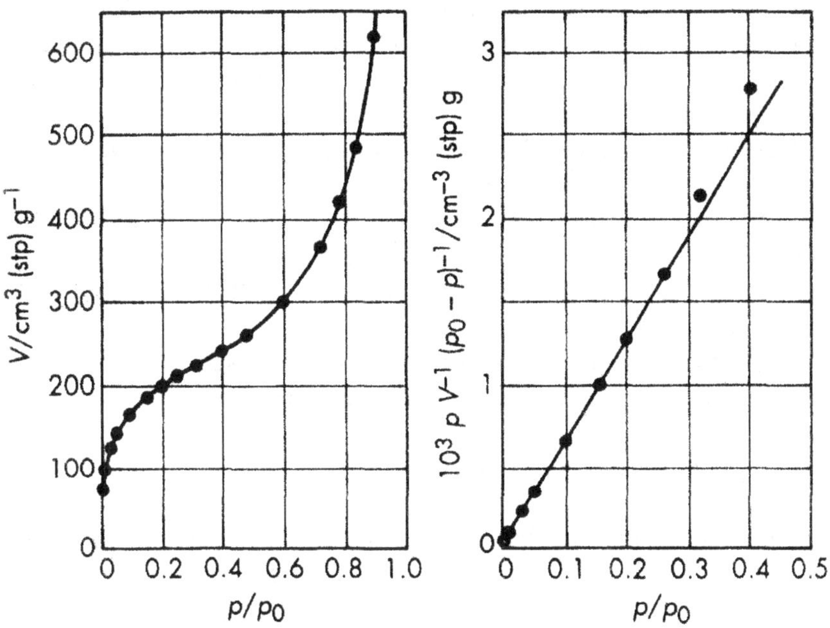

Figure 5.13 Isotherm and BET plot for the multilayer adsorption of nitrogen on a non-porous sample of silica gel at 77 K

effective area occupied by each adsorbed nitrogen molecule at monolayer capacity can be calculated from the density of liquid nitrogen (0.81 g cm^{-3}) on the basis of a model of close-packed spheres. The value so obtained is $16.2 \times 10^{-20}\ m^2$. This calculation would seem to be oversimplified, but the effects of the approximations involved are largely self-cancelling[170]. Adoption of this value leads to BET areas for non-porous solids which are generally in accord with corresponding surface areas determined by other techniques.

With most other adsorbates, adjusted molecular areas (determined, for example, by calibration with nitrogen adsorption data) which are usually in excess of those calculated from liquid densities must be used, and, moreover, for a given adsorbate, this adjusted value usually varies from solid to solid. This is due mainly to a certain amount of localisation of the adsorbed gas in the first monomolecular layer with respect to the variously distributed potential energy minima of the solid lattices. To avoid excessive localisation, a low value of $-\Delta H_1$ is desirable; however, a high value of c (i.e. $-\Delta H_1 \gg -\Delta H_L$) is also required to give a well-defined point B. The main reason why nitrogen is a particularly suitable adsorbate for surface area determinations is because the value of *c* is generally high enough to give a well-defined point B, but not too high to give excessive localisation of adsorption.

Effective cross-sectional areas of molecules adsorbed on to solid surfaces are catalogued in reference [72].

Krypton adsorption at 77 K is often used for the determination of relatively low solid-surface areas. At this temperature the vapour pressure of krypton (and so the dead-space correction) is small, and a reasonable precision is attainable.

The specific surface area of a powder can be used to estimate the average particle size if the particle shape is known.

Thermal desorption spectroscopy

A clean solid surface at fixed temperature is exposed to gas at low pressure (*c.* 10^{-8}–10^{-7} torr) to allow an appropriate amount of chemisorption to take place and the remaining adsorptive is then pumped off. The sample is heated rapidly at a programmed rate. Pressure surges are detected in characteristic temperature regions at which the rate of chemidesorption from the various types of adsorption site becomes appreciable. In addition to allowing different types of chemisorption site to be distinguished, this technique provides a means for estimating activation energies of desorption, which, in turn, set upper limits for the corresponding adsorption energies.

Composition and structure of solid surfaces

So far in this chapter, gross properties of solid surfaces (specific surface area, pore size distribution and adsorption energies) have been considered. In Chapter 3, the use of the electron microscope to investigate surface morphology was discussed. In this section, some of the spectroscopic, diffractometric and other techniques which are available for investigating the atomic detail at solid surfaces will be considered. Most of these are fairly modern and exploit high technology to the full (at corresponding cost!). Like some of the techniques referred to in Chapter 3 (TEM, SEM, PCS, etc.), they are mostly known by acronyms. They are many in number (Adamson[2] quotes over twenty (not counting minor variations)), but only a few of the more important techniques have been singled out for discussion here.

In addition to their particular instrumental requirements, the viability of most of these techniques is due to advances in high vacuum technology. In order that clean surfaces may be examined, they must be presented in a suitably uncontaminated state (see page 119) and recontamination during the course of the experiment must be avoided. Conventional high vacuum equipment (diffusion pumps) permits evacuation down to about 10^{-6} torr and is used to remove most of the bulk gas. However, even at this pressure, of the order of ten of the remaining gas molecules will collide with each surface atom every second, so, with only 1 per cent of these molecules actually sticking, the solid surface would be well contaminated within ten seconds. It is, therefore, often necessary to achieve ultra-high vacuum (UHV) conditions down to about 10^{-10} torr. There are several ways of doing this, but the pumped gas remains in the system rather than being removed. Ion pumps use electrons to ionise the gas molecules, which are collected and remain adsorbed at a metallic cathode. Cryopumps use very low temperature to promote gas removal by adsorption. Ionisation gauges are used to measure the UHV pressure thus achieved. Stainless steel and aluminosilicate glass are the materials favoured for apparatus construction and a malleable metal, such as gold, is used to obtain leak-free valves.

Having considered the preservation of a clean solid surface during the course of its study, some further comment on the preparation of the solid sample in the first place is appropriate. Some cleaning and polishing of the solid prior to its introduction into the UHV system may be necessary. Various subsequent treatments are then possible. Heating will help to remove chemisorbed gas, but introduces the risk of sintering. Ion bombardment (Ar^+ or electrons) or the use of pulsed laser light is also effective in removing chemisorbed gas, but can cause structural damage and leave the solid surface pitted. The researcher is, to some extent, faced with the dilemma that no treatment means a contaminated surface, but treatment means a modified surface. The solid surface to be studied is often generated *in situ*. Metal films can be produced by evaporating metal from an electrically heated filament. By means of remote control manipulation, a solid sample can be crushed under UHV to expose a large area of fresh surface. In order to obtain far more detailed information, solids can be cleaved to expose specific crystal faces. A study of clean solid surfaces, either as such or following exposure to known adsorbate, is then possible.

The most obvious application of these studies is towards an understanding of heterogeneous catalysis. Despite its immense importance, heterogeneous catalysis has tended to be an empirical subject, with the mode(s) of action far from understood. With the advent of modern surface chemical techniques, this situation is changing, though not without certain difficulties. The information provided can be of controversial value for a number of reasons. Usually, only stable species are detected and these may be of limited significance when considering the overall reaction mechanism. The surface may be unrepresentative by virtue of its mode of preparation and/or the perturbation caused by the experiment itself. The experiments involve low pressures, whereas industrial heterogeneous catalytic processes tend to be at atmospheric, or higher, pressure. The consequent 'pressure gap', therefore, involves a factor of the order of somewhere between about 10^8 and 10^{14}!

The two fundamental questions concerning heterogeneous catalysis relate to the surface chemistry and to the kinetics of the process, and are, respectively,

1. what is the nature of the catalyst surface?
2. what is the reaction mechanism?

These questions, of course, are strongly related. The following surface chemical techniques are described and exemplified very much with the first of these questions in mind, including subsidiary questions such as to (*a*) the nature of an active site and (*b*) the role of catalyst promoters. For an account of heterogeneous catalytic reaction mechanisms, the reader is referred to modern texts on reaction kinetics/catalysis.

Electron spectroscopy

Electron spectroscopy is the study of electrons emitted when matter is irradiated with photons or bombarded with particles. In *photoelectron spectroscopy* (PES), a monochromatic source of irradiation is used and the kinetic energies of directly ejected (primary) electrons are analysed by means of an electron spectrometer, i.e. an intensity versus energy spectrum is produced. This kinetic energy, E_{el}, is given in terms of the irradiation frequency, ν, and the ionisation potential, I, of the electron by Einstein's equation

$$E_{el} = h\nu - I \tag{5.14}$$

If it is assumed that photoionisation occurs without any adjustment of the remaining electrons (Koopman's theorem), the ionisation potential and the orbital energy, ϵ, of the ejected electron are numerically equal ($I = -\epsilon$). A spectrum of electron orbital energies is thus obtained.

The ejected electrons will interact strongly with matter because of their charge, therefore, when studying solids, the detection of electrons as described above will be limited to those ejected from atoms at or close to the surface.

In *ultraviolet photoelectron spectroscopy* (UPS or U-PES), the irradiation (usually a He(I) (21.2 eV) or He(II) (40.8 eV) source) causes the displacement of a valence electron. Although an important method of studying the electronic nature of molecules in the gas phase, it is less useful for studying the surfaces of metals, since the valence electrons are in a continuous (conducting) energy band with a spread of about 10 eV. Adsorbed layers can, however, usefully be investigated in terms of the difference between the spectrum following adsorption and that for the clean metal surface.

In *X-ray photoelectron spectroscopy* (XPS or X-PES), the irradiation (usually a Mg K α (1253.6 eV) or Al K α (1486.6 eV) source) causes a core electron to be ejected. This is a more useful technique than UPS for surface studies, since the binding energies of core electrons are characteristic of the elements in question and surface elements can thus be identified by the traditional 'spectroscopic fingerprinting' procedure. In this respect, XPS is sometimes referred to by its alternative name, *electron spectroscopy for chemical analysis* (ESCA). In order to emphasis the contribution from surface atoms, the X-ray beam is usually set at a grazing angle to the surface. Most of the signal originates from within a nanometre of the surface.

The ionisation potential of a core electron depends, to a small extent, on the chemical environment of the atom in question, and chemical shifts of up to about 10 eV can be observed. For example, the C(1s) XPS signal for molecularly adsorbed carbon monoxide on polycrystalline iron at 290 K shows a peak at 285.5 eV, which is gradually replaced by a peak at 283 eV on heating to 350 K as a result of surface dissociation[73]. Likewise, chemical shifts can be exploited to study the oxidation states of surface atoms.

Table 5.1 Mole percentage compositions from XPS study of the promoted iron catalyst used in ammonia synthesis[171,172]

	Fe	*Al*	*K*	*Ca*	*O*
Bulk	41	2	0.4	1.7	53
Surface (unreduced)	3	8	33	4	51
Surface (reduced)	6	16	24	4	49

Table 5.1 shows an application of XPS to the study of the promoted iron catalyst used in the Haber synthesis of ammonia. The sizes of the various electron intensity peaks allows a modest level of quantitative analysis. This catalyst is prepared by sintering an iron oxide, such as magnetite (Fe_3O_4) with small amounts of potassium nitrate, calcium carbonate, aluminium oxide and other trace elements at about 1900 K. The unreduced solid produced on cooling is a mixture of oxides. On exposure to the nitrogen–hydrogen reactant gas mixture in the Haber process, the catalyst is converted to its operative, reduced form containing metallic iron. As shown in Table 5.1, the elemental components of the catalyst exhibit surface enrichment or depletion, and the extent of this differs between unreduced and reduced forms.

Following the ejection of an inner electron (as in XPS), an electron from a higher orbital will fall into the vacant orbital. The energy so released may be emitted as radiation (X-ray fluorescence), or it may cause another electron from a higher orbital (a secondary, or Auger, electron) to be ejected. Auger emission is favoured from light elements and X-ray fluorescence from heavy elements. Hydrogen and helium will, of course, not give an Auger signal, since they have no L-shell electron. Lithium, with only one L-shell electron would seem unable to give an Auger signal, but it does so by means of a cooperative effect between two atoms. Owing to the consequent low energy of the Auger electrons, *Auger electron spectroscopy* (AES) of solids is more biased in favour of the surface atoms than XPS. Each element has a characteristic AES spectrum which can be used for qualitative and semiquantitative analysis.

Suppose that a primary 1s electron is ejected, following which a 2s electron falls into the vacant orbital and a 2p Auger electron is ejected. Labelling these electrons K, L1 and L2, respectively, this is a KLL-type emission and the energy of the resulting Auger electron is given by

$$E_{\text{Auger}} = I(\text{K}) - I(\text{L1}) - I(\text{L2}) \tag{5.15}$$

Since this is a function of ionisation energies only, it is not necessary to use a monochromatic energy source, which is an attractive experimental feature of the technique. Indeed, Auger electrons can be identified as such by their independence of the incident energy. With many experimental features in common, it is possible to construct a single instrument with facilities for XPS, AES, LEED and certain other studies. The energy source for AES is usually an electron beam, rather than X-rays. The disadvantage of an electron beam is that it may cause structural damage. The advantage is that it can be focussed (as in the electron microscope) and scanned over the surface – *scanning Auger electron spectroscopy (SAES)*. It is possible to focus down to 10^{-14} m^2 and systematically scan an area of 10^{-8} m^2. A facility for sputter-ion etching by Ar^+ ion bombardment can be included. This enables the surface layers to be removed progressively prior to studying the underlying surface features. AES is similar, in principle, to SEM (see page 49) except that the latter employs an incident electron beam of higher energy.

AES can be sensitive to as little as 1 per cent of a monolayer of adsorbed material and is often used as a preliminary investigation of the cleanliness of a surface prior to its study by LEED.

Figure 5.14 shows SAES analysis of surface iron, aluminium, calcium and potassium on reduced ammonia synthesis catalyst. As can be seen, these surface distributions are patchy rather than uniform. The calcium and, to a lesser extent, the aluminium tend to be located in areas of their own where iron is absent, but the potassium tends to be located alongside the iron. This supports the view that calcium and aluminium oxides are structural promoters, with the role of keeping the iron crystallites apart and minimising loss of surface area through sintering. Potassium is probably a chemical promoter, i.e. the positively charged potassium ions influence the electronic structure in the nearby iron such as to give enhanced catalytic activity.

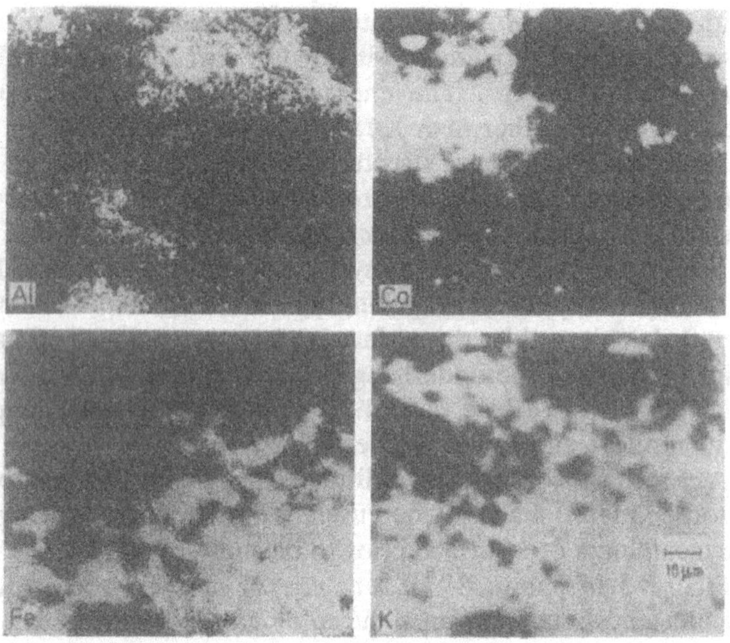

Figure 5.14 SAES surface analysis of Fe, Al, Ca and K on reduced promoted iron catalyst used for ammonia synthesis[74, 171]
(*By courtesy of Springer-Verlag, Heidelberg*)

AES can be used for studying specific crystal faces. For example, nitrogen chemisorption on iron has been shown to be much stronger on the more open Fe(111) crystal faces than on the more compact Fe(100) and Fe(110) faces (see Figure 5.16).

Low energy electron diffraction (LEED)

We now turn from the study of surface composition to that of surface structure. Low energy electron diffraction is the basic technique for studying the arrangement of atoms at solid surfaces, just as X-ray diffraction is the basic technique for studying the three-dimensional arrangements of atoms in crystalline solids. The electron energy range exploited in LEED is roughly 20–300 eV. In this range, two basic, but conflicting, requirements are satisfied.

1. The electron energy is high enough for the de Broglie wavelength ($\lambda = h/p = h/(2m\,\epsilon)^{1/2}$) to be of the same order of magnitude as interatomic spacings (e.g. ϵ = 150 eV corresponds to $\lambda = 10^{-10}$ m).
2. The energy is low enough to ensure that most of the electrons (incident at right-angles to the surface) do not penetrate much beyond the outer atomic layer of the solid, thus giving a technique appropriate for surface studies.

Crystalline solids consist of periodically repeating arrays of atoms, ions or molecules. Many catalytic metals adopt *cubic close-packed* (also called *face-centred cubic*) (Co, Ni, Cu, Pd, Ag, Pt) or *hexagonal close-packed* (Ti, Co, Zn) structures. Others (e.g. Fe, W) adopt the slightly less efficiently packed *body-centred cubic* structure. The different crystal faces which are possible are conveniently described in terms of their *Miller indices*. It is customary to describe the geometry of a crystal in terms of its *unit cell*. This is a parallelepiped of characteristic shape which generates the *crystal lattice* when many of them are packed together.

In two dimensions, the equivalents of unit cell and lattice are *unit mesh* and *net*, respectively. Crystallography in two dimensions is, obviously, simpler than that in three dimensions, and there are only five types of net (illustrated in Figure 5.15). The choice of unit mesh is arbitrary. The primitive unit mesh (illustrated at the bottom left hand corner of each net) is the smallest possible repeating quadrilateral with lattice points only at the corners. However, it may be appropriate

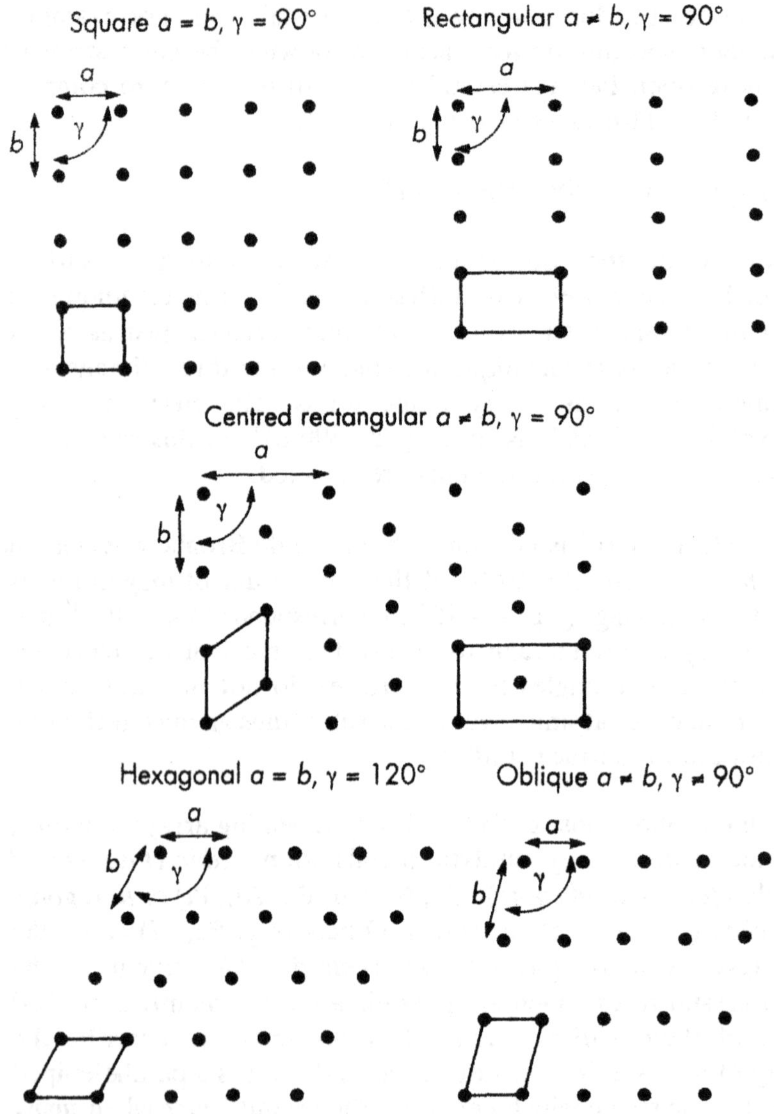

Figure 5.15 Classification of nets

to make a different choice of unit mesh in order to show the symmetry of the system more clearly. For. example, the centred rectangular net can, alternatively, be viewed as an oblique net, as the primitive mesh illustrates; however, the choice of a centred rectangular unit mesh (bottom right hand corner) illustrates the surface geometry better.

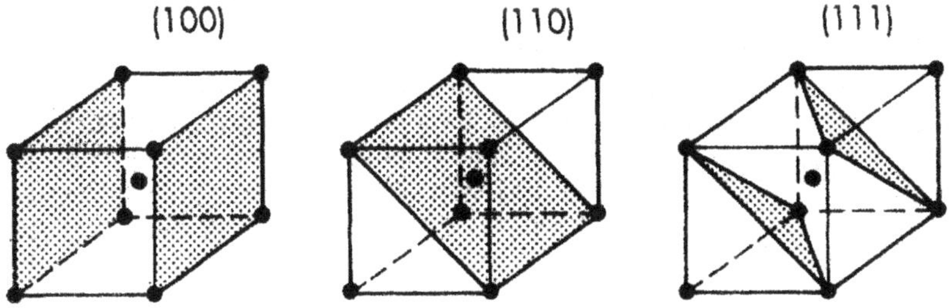

Figure 5.16 Illustration of the (100), (110), and (111) planes of a body-centred cubic lattice (e.g. Fe). For clarity, the (100) and (111) planes through the centre atom are not shown. As can be seen, the nets are square, centred rectangular and hexagonal, respectively

The structure of a surface layer (be it the surface of a pure solid or a monolayer of adsorbed gas) usually differs from that of the underlying substrate. A shorthand notation exists in which the unit mesh of the surface layer is described in terms of the unit mesh of the layer immediately below it. Some examples are illustrated in Figure 5.17. The prefix 'C' indicates that the unit mesh for the surface layer contains centre atom(s) and 'R' indicates that this unit mesh is rotated by the stated angle with respect to the substrate unit mesh.

In examples (*a*) to (*c*), the locations of the adsorbed atoms differ in that they are (*a*) end-on, (*b*) in a two-fold bridging position, and (*c*) in a three-fold well position in relation to the substrate atoms. However, the outer surface net is the same in each case and, as such, contributes to the same diffraction pattern. In example (*d*), the alternative unit meshes shown are both correct and both would lead to the same ultimate interpretation, but the C(2 × 2) unit mesh offers the greater convenience.

Figure 5.18 shows schematically an experimental arrangement for LEED studies. Electrons from a heated metal cathode are given uniform acceleration and strike the crystal face under investigation normal to its surface. The resulting electron scattering may be elastic (diffraction) or inelastic. Negatively charged grids (G1) are set at potentials such that only the elastically scattered electrons have sufficient energy to pass. These are then accelerated by a positively charged grid (G2) and the diffraction pattern is displayed on a fluorescent screen. Ultra-high vacuum must be maintained to avoid surface contamination and secondary scattering. Chemisorbed layers are studied on the basis that they will remain intact when the adsorptive has been pumped away.

Consider the diffraction from a one-dimensional row of regularly spaced atoms. From Figure 5.19 it can be seen that the condition for constructive interference is

$$\Delta = d \sin \theta = n\lambda \tag{5.16}$$

The diffracted maxima will, therefore, lie on the surfaces of cones, as illustrated in Figure 5.20.

Diffraction from the rows of atoms in the second direction will be similar and the diffraction maxima for the two-dimensional net will be an array of points corresponding to the lines where the maxima cones for each dimension meet. From equation (5.16) it can be seen that the spacings between diffraction maxima are inversely propor-

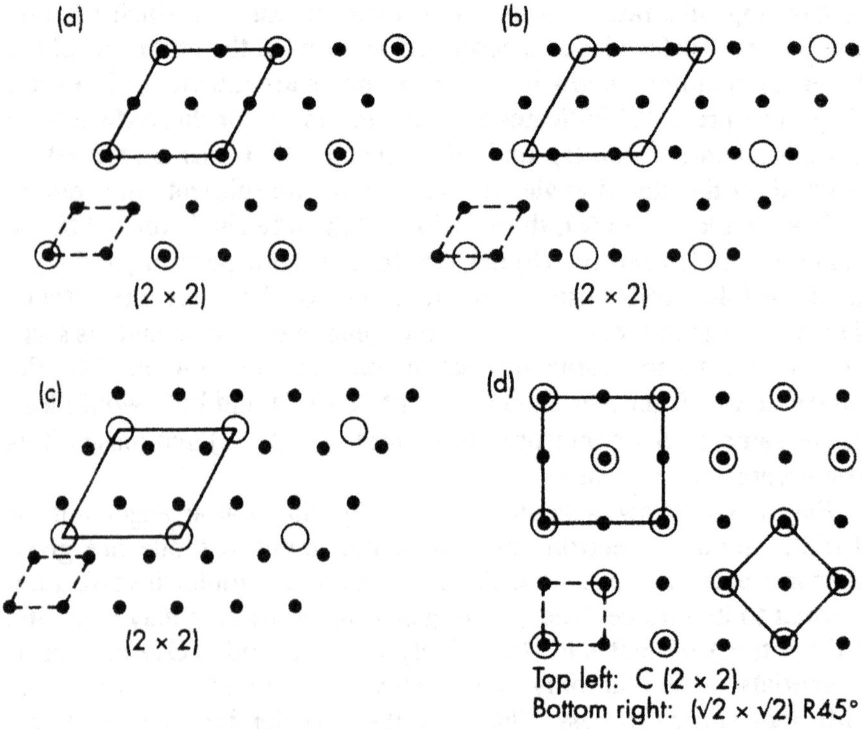

Figure 5.17 Examples and nomenclature of surface layers. Substrate atoms are represented by dots and adatoms by circles. The unit (1×1) mesh of the substrate is shown bottom left

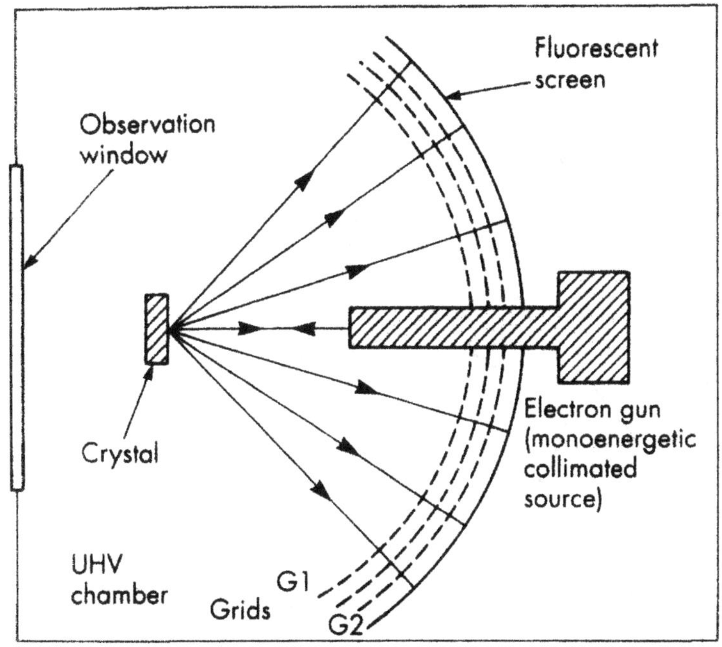

Figure 5.18 Schematic illustration of a LEED apparatus

tional to the spacings between the scattering sources. The resulting LEED pattern (as with any other diffraction pattern), therefore, shows the surface net in 'reciprocal space' rather than directly, as illustrated in Figure 5.21.

The C (2 × 2) surface layer shown in Figure 5.21b is, in effect, two slightly displaced (2 × 2) surface layers, the first as shown and the

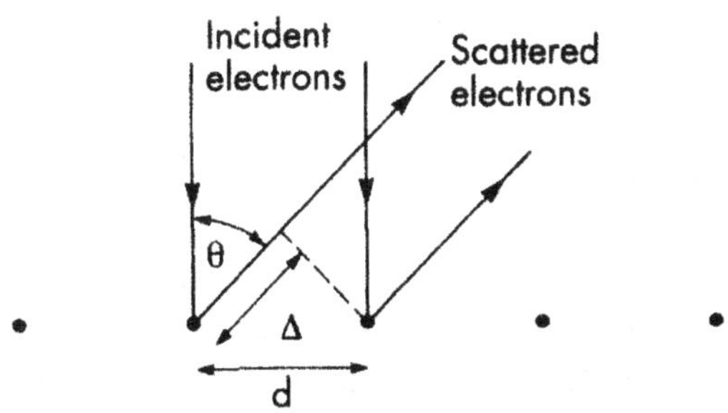

Figure 5.19 Diffraction by a row of atoms for normal incidence

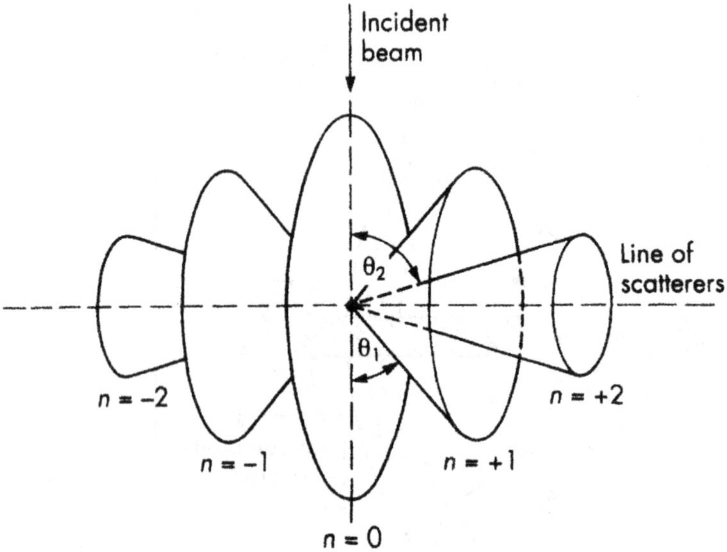

Figure 5.20 Cones of diffracted radiation from a one-dimensional row of point scatterers

second comprising the 'centre atoms'. The resulting diffraction pattern is, therefore, that of a (2 × 2) surface layer in which the spacing is half of that which would be given by the underlying substrate atoms. Since the immediate substrate atoms will in practice give some (albeit attenuated) signal, the LEED pattern will consist of an array of bright (surface layer only) and very bright (surface layer plus substrate) spots, as illustrated.

Suppose now that only a proportion of these surface layer sites are occupied by adsorbate. The consequent combination of order and randomness in the surface layer will give a LEED pattern which shows the same spots (but at reduced intensity) together with some diffuseness around the centre spot.

Field emission and field ionisation microscopy

In *field emission spectroscopy* (FEM), a refactory metal, such as tungsten, is fabricated to give a very fine hemispherical tip of radius of curvature about 10^{-7} m. The tip is located at the centre of curvature of a hemispherical fluorescent screen and a potential difference of about 10 kV is applied, with the fluorescent screen as the anode. Electrodes are wrenched from the metal tip, accelerated along radial lines and hit the fluorescent screen where they display a much magnified (*c*. 10^6) picture of the distribution of surface atoms in the metal tip.

In *field ionisation microscopy* (FIM), helium at low pressure is introduced into the above system and the polarity of the applied potential difference is reversed. Helium atoms in the vicinity of the now positively charged metal tip are stripped of an electron and the resulting helium ions are accelerated radially to the negatively charged fluorescent screen.

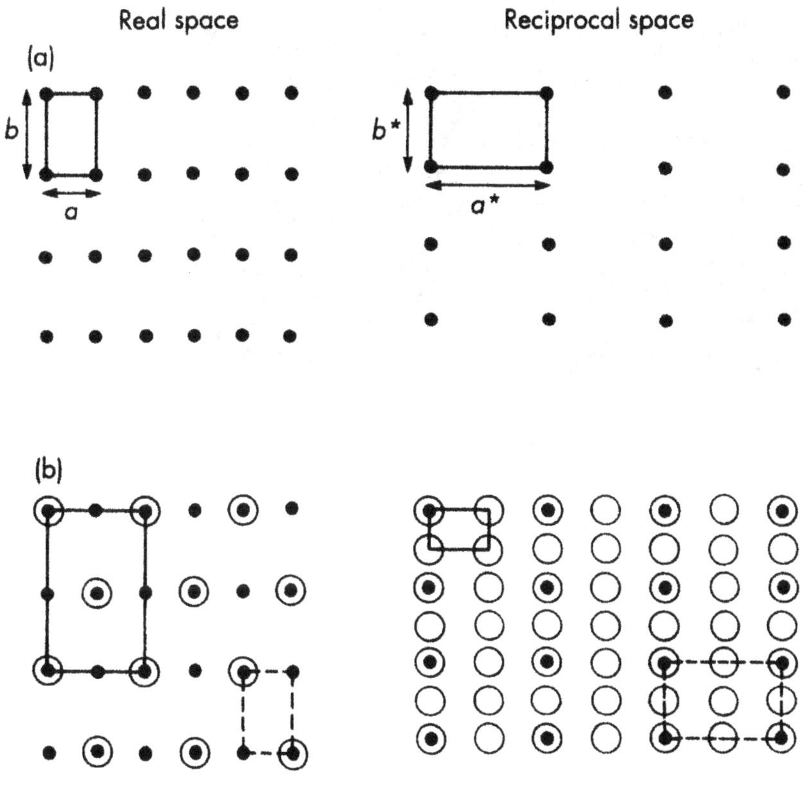

Figure 5.21 Real and reciprocal space. Substrate atoms are represented by dots and adatoms by circles. The relationship, $a/b = b^*/a^*$, holds, but the absolute dimensions of the reciprocal space mesh, of course, depend on instrumental magnification

Figure 5.22 FIM image of the surface atoms at a tungsten Tip[66] (*By courtesy of Oxford University Press*)

6 The solid–liquid interface

Contact angles and wetting[75–77]

Wetting is the displacement from a surface of one fluid by another. It involves, therefore, three phases, at least two of which must be fluids. The following account will be restricted to wetting in which a gas (usually air) is displaced by a liquid at the surface of a solid. A wetting agent is a (surface-active) substance which promotes this effect.

Three types of wetting can be distinguished:

1. Spreading wetting.
2. Adhesional wetting.
3. Immersional wetting.

Spreading wetting

In spreading wetting, a liquid already in contact with the solid spreads so as to increase the solid–liquid and liquid–gas interfacial areas and decrease the solid–gas interfacial area. The spreading coefficient, S (cf. equation 4.31), is defined by the expression

$$\begin{aligned} S &= -\Delta G_s / A \\ &= \gamma_{SG} - (\gamma_{SL} + \gamma_{LG}) \end{aligned} \tag{6.1}$$

where ΔG_S is the free energy increase due to spreading. The liquid spreads spontaneously over the solid surface when S is positive or zero.

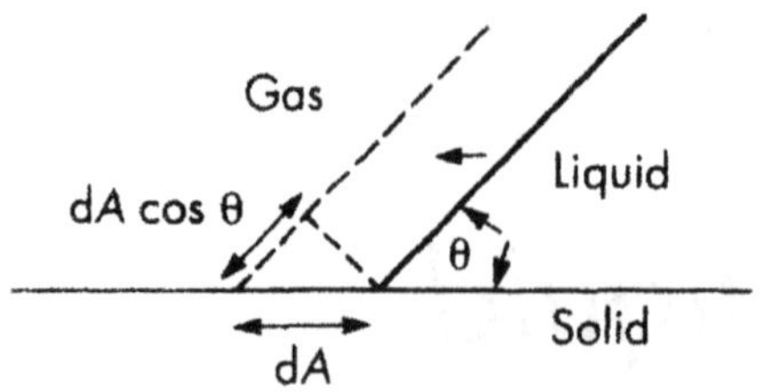

Figure 6.1

When S is negative, the liquid remains as a drop having a definite angle of contact, θ, with the solid surface. The equilibrium contact angle is such that the total surface free energy of the system is a minimum – i.e. $\gamma_{SG} A_{SG} + \gamma_{SL} A_{SL} + \gamma_{LG} A_{LG}$ is a minimum, where A represents interfacial area. Consider a liquid making an equilibrium contact angle, θ, to spread an infinitesimal amount further so as to cover an extra area, dA, of the solid surface. The increase in liquid–gas interfacial area is, therefore, dA cos θ (see Figure 6.1) and the increase in the free energy of the system is given by

$$\mathrm{d}G = \gamma_{SL}\mathrm{d}A + \gamma_{LG}\mathrm{d}A \cos\theta - \gamma_{SG}\mathrm{d}A$$

If the system is at equilibrium, dG = 0, and

$$\gamma_{SL} + \gamma_{LG} \cos\theta - \gamma_{SG} = 0 \tag{6.2}$$

In this expression (known as Young's equation), γ_{SG} is the surface tension of the solid in equilibrium with the vapour of the wetting liquid. If γ_S is the surface tension of the solid against its own vapour, then

$$\gamma_S - \gamma_{SG} = \pi_{SG}$$

and

$$\gamma_{SL} - \gamma_S + \gamma_{LG} \cos\theta + \pi_{SG} = 0 \tag{6.3}$$

where π_{SG} (the spreading pressure) is the reduction of the surface tension of the solid due to vapour adsorption. In general, π_{SG} is small for moderately large values of θ (and equation (6.2) applies), but it can become significant as θ approaches zero[75].

If Fowkes' semiempirical interfacial tension theory (as described on pages 65–67) is applied to the solid–liquid interface, then

$$\gamma_{SL} = \gamma_S + \gamma_{LG} - 2(\gamma_S^d \times \gamma_{LG}^d)^{1/2} \tag{6.4}$$

which, combined with equation (6.2), gives

$$\cos\theta = -1 + \frac{2(\gamma_S^d \times \gamma_{LG}^d)^{1/2}}{\gamma_{LG}} \tag{6.5}$$

or, for non-polar liquids, where $\gamma_{LG}^d = \gamma_{LG}$,

$$\cos\theta = -1 + 2(\gamma_S^d / \gamma_{LG})^{1/2} \tag{6.6}$$

For a selection of non-polar liquids on a given solid, it follows that θ should decrease as γ_{LG} decreases and become zero below a certain value of γ_{LG}. Zisman[78] has named this value of γ_{LG} the *critical surface tension*, γ_c, for the solid. Critical surface tension is a useful parameter for characterising the wettability of solid surfaces (see Table 6.1).

Table 6.1 Critical surface tensions for solid surfaces (After Zisman[78])

Solid surface	γ_c(20°C)/mN m^{-1}
Condensed monolayer with close-packed terminal CF_3 groups	6
Polytetrafluoroethylene	18
Polytrifluoroethylene	22
Poly(vinylidine fluoride)	25
Poly(vinyl fluoride)	28
Polyethylene	31
Polystyrene	33
Poly(vinyl alcohol)	37
Poly(vinyl chloride)	40
Poly(hexamethylene adipamide) (nylon 66)	46

To determine γ_c for a solid, the advancing contact angles made by several non-polar or weakly polar liquids are measured and the value of γ_{LG} which corresponds to $\theta = 0$ (i.e. $\cos\theta = 1$) is found by graphical extrapolation. Zisman has determined γ_c for many solid surfaces by way of empirical plots of $\cos\theta$ versus γ_{LG} (see Figure 6.2). According to equation (6.6), however, plots of $\cos\theta$ versus $(\gamma_{LG})^{-1/2}$ should be approximately linear and permit more reliable extrapolation to $\cos\theta = 1$.

Adhesional wetting

In adhesional wetting, a liquid which is not originally in contact with the solid substrate makes contact and adheres to it. In contrast to spreading wetting, the area of liquid–gas interface decreases. The work (free energy) of adhesion is given by the Dupré equation (see equation (4.29) in the form

$$
\begin{aligned}
W_a &= -\Delta G_a / A \\
&= \gamma_{SG} + \gamma_{LG} - \gamma_{SL}
\end{aligned} \tag{6.7}
$$

which, combined with Young's equation (6.2), gives the Young–Dupré equation,

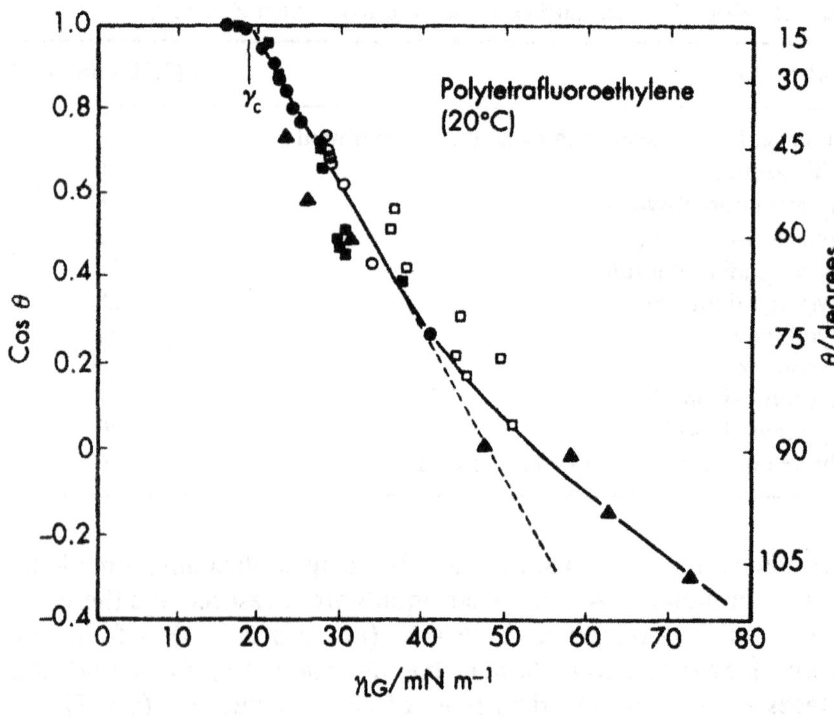

Figure 6.2 Critical surface tension of polytetrafluoroethylene[78]. ●, *n*-alkanes; ○, other hydrocarbons; ■, esters and ethers; □, halocarbons and halohydrocarbons; ▲, miscellaneous liquids

$$W_a = \gamma_{LG} (1 + \cos\theta) \tag{6.8}$$

For zero contact angle, $\cos\theta = 1$ and $W_a = 2\,\gamma_{LG} = W_c$; therefore, zero contact angle results when the forces of attraction between liquid and solid are equal to or greater than those between liquid and liquid, and a finite contact angle results when the liquid adheres to the solid less than it coheres to itself.

The solid is completely wetted by the liquid if the contact angle is zero and only partially wetted if the contact angle is finite. Complete non-wetting implies a contact angle of 180°, which is an unrealistic situation, since it requires that either $W_a = 0$ or $\gamma_{LG} = \infty$. There is always some solid–liquid attraction; for example, water droplets will adhere to the underside of a paraffin wax surface ($\theta \sim 110°$).

Immersional wetting

In immersional wetting, the solid, which is not originally in contact with the liquid, is immersed completely in the liquid. The area of liquid–gas interface, therefore, remains unchanged. The free energy change for immersion of a solid in a liquid is given by

$$\begin{aligned} -\Delta G_i &= \gamma_{SG} - \gamma_{SL} \\ &= \gamma_{LG} \cos\theta \end{aligned} \tag{6.9}$$

If $\gamma_{SG} > \gamma_{SL}$, then $\theta < 90°$ and immersional wetting is spontaneous, but if $\gamma_{SG} < \gamma_{SL}$, then $\theta > 90°$ and work must be done to immerse the solid in the liquid.

Free energy, enthalpy and entropy of immersion are related by

$$\Delta G_i = \Delta H_i - T\Delta S_i \tag{6.10}$$

ΔH_i can be measured directly by sensitive calorimetry. ΔG_i and ΔH_i can be equated only when ΔS_i is negligible.

Measurement of contact angles[79]

Given a flat solid surface, the actual measurement of contact angles to ±1° is relatively straightforward. The complications associated with contact angle measurement relate to the system itself and include the following:

1. Contamination of the liquid usually influences the contact angle.
2. Solid surfaces differ from liquid surfaces in that they show a far greater degree of heterogeneity, even after careful polishing; for example, a solid surface polished to the best optical standards is wavy and pitted compared with a quiescent liquid surface. To obtain a solid surface free from impurities which are likely to have a significant effect on its properties is usually very difficult. It can, therefore, be appreciated that any measured property of a solid surface is subject to variability as a result of unavoidable sample differences.
3. In practice, contact angles are rarely single-valued quantities, but, for a given system, a range of metastable contact angles exists. The observed contact angle will depend mainly on (*a*) whether the liquid is advancing over a dry surface or receding from a wet surface, and (*b*) the extent to which the drop is vibrated. Contact angle hysteresis is most noticeable with chemically and/or geometrically heterogeneous surfaces. The difference between advancing and receding contact angles may be as much as 50°. Consider the spreading of a liquid on a solid surface with a mixture of high energy (low θ) and low energy (high θ) regions. Advancing liquid will spread relatively easily over the high energy regions of the surface, but tend to stick and locate the triple interface at low energy regions, thus giving a high advancing contact angle. Receding liquid, on the other hand, will readily vacate low energy regions of the surface and the triple interface will tend to be located at high energy regions, thus giving a low receding contact angle. Contact angle hysteresis on geometrically rough surfaces can be considered, to some extent, in terms of a constant microscopic angle leading to different macroscopic angles according to the local inclination of the surface. A common example of contact angle hysteresis is given by raindrops on a dirty windowpane. Mercury rolls off glass, and other solids, very easily because of low contact angle hysteresis.
4. Liquid drops on anisotropic solid surfaces will tend to elongate in the higher surface energy direction and the contact angle will, therefore, vary with position.

If a moderately large area of flat solid surface is available, contact angles are usually measured directly from a projection of a sessile drop of the liquid. Alternatively, the tilting-plate method illustrated

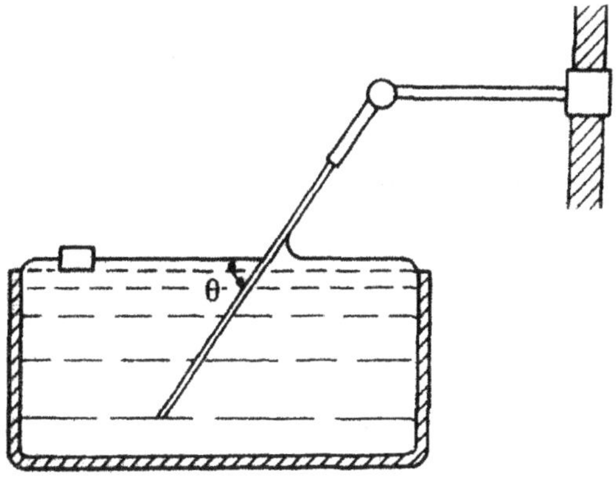

Figure 6.3 Tilting plate method for the measurement of contact angles

in Figure 6.3 can be used; the angle of the plate is adjusted so that the liquid surface remains perfectly flat right up to the solid surface. Another method, originally devised by Langmuir and Schaeffer, is based on observation of the angle at which light from a point source is reflected from the surface of a liquid drop at its contact point with a plane solid surface[173–174]. This technique can be improved by using laser-light. It has been refined for measuring contact angles formed by liquids on the surfaces of small-diameter filaments[174].

The contact angles of solids in a finely divided form are technically important (e.g. flotation; page 161) but are difficult to measure. Bartell *et al.*[175] developed a method based on displacement pressures for measuring such contact angles. The finely divided solid is packed into a tube and the resulting porous plug is considered to act as a bundle of capillaries of some average radius r. The pressure required to prevent the liquid in question from entering the capillaries of this porous plug is measured, and the equation for the pressure difference across a curved liquid surface (5.4) is applied,

$$p = 2\gamma \cos \theta / r \qquad (6.11)$$

The equivalent radius of the capillaries is then found from a similar experiment using a liquid which completely wets the solid – i.e.

$$p' = 2\gamma' / r \qquad (6.12)$$

In a related method[176] the contact angle is determined by measuring the rate of penetration of liquid through a powder packed into a glass tube.

Factors affecting contact angles and wetting

The three types of wetting are summarised by the following equations:

$$-\Delta G_{\text{spreading}} / A = S = \gamma_{SG} - \gamma_{SL} - \gamma_{LG} \tag{6.1}$$

$$-\Delta G_{\text{adhesion}} / A = W_a = \gamma_{SG} - \gamma_{SL} + \gamma_{LG} \tag{6.7}$$

$$-\Delta G_{\text{immersion}} / A = \gamma_{SG} - \gamma_{SL} = \gamma_{LG} \cos\theta \tag{6.9}$$

A reduction of γ_{SL} facilitates all of these wetting processes, but a reduction of γ_{LG} is not always helpful.

The contact angle between water and glass is increased considerably by even less than an adsorbed monolayer of greasy material such as fatty acid. W_a is decreased, since some of the glass–water interface is replaced by hydrocarbon–water interface (Figure 6.4a); hence, from the Young–Dupré equation, θ increases.

Figure 6.4

The wetting of a hydrophobic solid surface by an aqueous medium is considerably helped by the addition of surface-active agents. W_a is increased and γ_{LG} is decreased (Figure 6.4b), so that, from the Young–Dupré equation, θ is reduced on two counts.

Surface roughness has the effect of making the contact angle further removed from 90°. If θ is less than 90°, the liquid will penetrate and fill up most of the hollows and pores in the solid and so form a plane surface which is effectively part solid and part liquid; since liquid has zero θ with liquid, θ will, therefore, decrease. On the other hand, if θ is greater than 90°, the liquid will tend not to penetrate into the hollows and pores in the solid and can, therefore, be regarded as resting on a plane surface which is effectively part solid and part air; since there is practically no adhesion between the liquid and the entrapped air, θ will increase. Surface roughness is a possible cause of contact angle hysteresis on this basis.

The method used to prepare the solid surface may affect the contact angle; for example, substances which have crystallised in contact with water often have a lower contact angle than if they crystallise in air, owing to the orientation of the water-attracting groups outwards in the former case. Penetration and entrapment of traces of water in the surface layers, which decreases θ, is also probable in these circumstances.

Fluorocarbon surfaces have characteristically low critical surface tensions (see Table 6.1) and have found well-known application in the production of 'non-stick' surfaces. It is probable that fluorocarbon surfaces show much more pronounced non-wetting characteristics than the corresponding hydrocarbon surfaces, mainly on account of the large size of the $-CF_2-$ groups compared with that of the $-CH_2-$ groups. Since fewer $-CF_2-$ groups than $-CH_2-$ groups can be packed into a given area of the solid surface, W_a is less and θ is greater for the fluorocarbon surface[75].

A major difference between the wetting of 'hard' (e.g. glass and metal) and 'soft' (e.g. textile) solid surfaces is that, in the former, equilibrium tends to be established rapidly, whereas, in the latter, kinetic effects may be of considerable importance.

Wetting agents

Surface-active materials, particularly anionics, are used as wetting agents in many practical situations. For example, in dips for sheep and cattle and in the application of insecticide and horticultural sprays, the surfaces in question tend to be greasy or wax-like, thus presenting unfavourable conditions for satisfactory surface coverage unless a wetting agent is incorporated. However, in these cases complete wetting is not desirable either, since it causes over-efficiency in the drainage of excess liquid from the surface. Wetting agents also find considerable application in the textile industry for the purpose of obtaining even results in operations such as scouring, bleaching, mercerising and dyeing.

Cationic surfactants can be exploited to promote oil-wetting in processes such as dry-cleaning and road making.

In addition to lowering γ_{LG}, it is important that the wetting agent lowers γ_{SL}, so that a choice of surfactant to suit the particular nature of the solid surface has to be made (side-effects such as toxicity, foaming, etc., must also be borne in mind). Irregularly shaped surfactant molecules, e.g. sodium di-*n*-octyl sulphosuccinate (Aerosol OT), are often very good wetting agents, since micelle formation is not favoured, owing to steric considerations; this permits relatively high concentrations of unassociated surfactant molecules and, hence, greater lowering of γ_{LG} and γ_{SL}. Non-ionic surfactants are also good wetting agents.

Water repellency

This is the converse of the previous topic, the aim being to make the contact angle as large as possible. Textile fabrics are made water-repellent by treatment with a long-chain cationic surfactant (e.g. stear-amidomethylpyridinium chloride, $C_{17}H_{35}CO\ NH\ CH_2N^+C_5H_5\ Cl^-$).

A condition of negative capillary action is achieved. The pressure required to force water through the fabric depends on the surface tension and inversely on the fibre spacing, so that a moderately tight weave is desirable. The passage of air through the fabric is not hindered.

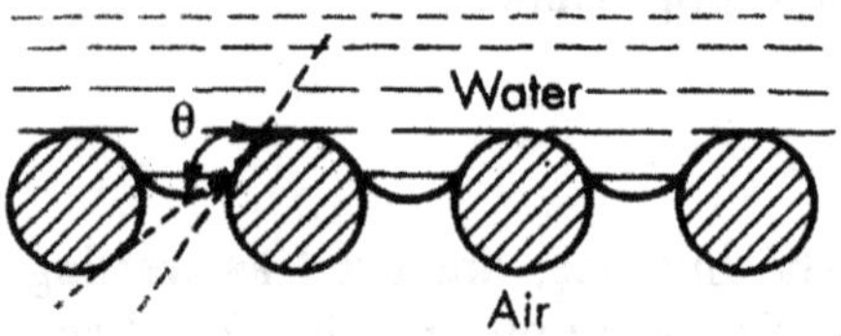

Figure 6.5

Ducks owe their water-repellent characteristics to the nature of their feathers, which consist of fine, wax-covered barbules *c.* 8 μm in diameter, separated by air gaps of *c.* 30 μm.

Dimethyldichlorosilane is a very good hydrophobising agent for silica and glass surfaces; it reacts with the –OH groups on the outside of the silicate lattice with the elimination of HCl to give

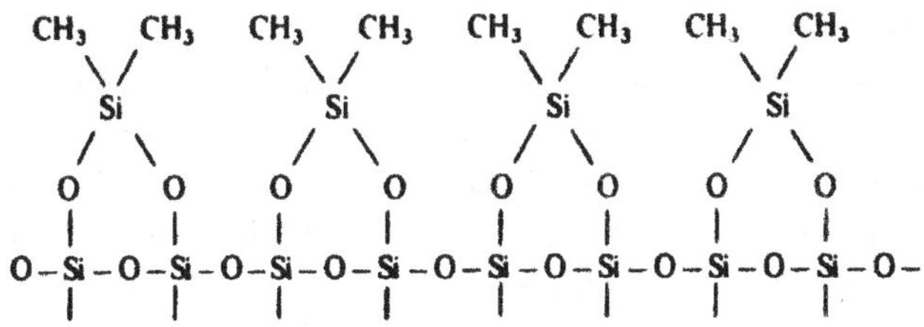

Ore flotation[80–81]

For a solid particle to float on the surface of a liquid, the total upward pull of the meniscus around it must balance the apparent weight of the particle; for example, a waxed needle can be floated on the surface of water (Figure 6.6) and then sunk by the addition of detergent. The flotation of a solid on a liquid surface depends on the contact angle θ, and since θ can readily be modified by factors such as surface grease, surfactants, etc., the conditions of flotation can also be controlled.

The various constituents of many crude ores have different tendencies to float on the surface of water and these tendencies can be modified advantageously by means of additives.

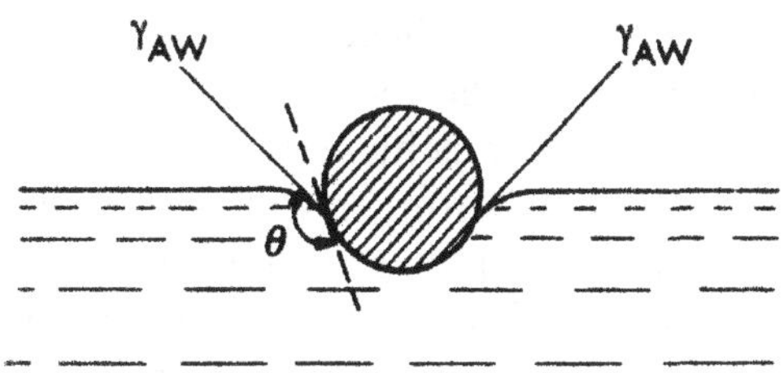

Figure 6.6

The world-wide treatment of crude ore by flotation amounts to *c.* 10^9 tonne per year. By concentrating metal ores on site, considerable saving of transportation costs can be effected, thus permitting the exploitation of low-grade ores (as little as 1 per cent metal content) which would otherwise be uneconomic.

After mining, the crude ore is crushed and then ground into a slurry in water (with particle diameters typically in the range 0.01–0.1 mm). A small quantity of collector oil is added during the grinding stage. The collector oil adsorbs strongly on the surfaces of the metal ore particles, with the result that the contact angle at the solid–liquid–air boundary increases to the point where flotation is possible. The collector oil does not adsorb so strongly on siliceous material, which remains wetted by water and does not float.

The collector oil molecules are amphiphilic, with their polar groups exhibiting some affinity for particular metal ores, and so create a hydrophobic particle surface on adsorption. They can be anionic, cationic or non-ionic. Organic xanthates and thiophosphates are often used for sulphide ores and long-chain fatty acids for oxide and carbonate ores.

A foaming agent, such as crude cresol or pine oil (soap is unsuitable, as it lowers θ too much), is added to the suspension of ground ore and collector oil in water and the pH is adjusted to give the particles low zeta potentials and, therefore, minimise electrostatic repulsions. Air is forced through a fine sieve at the bottom of the vessel. The particles of metal ore become attached to the air bubbles, which carry them to the surface (Figure 6.7), where they collect as a metal-rich foam which can be skimmed off.

Contact angles of at least 50–75° are required for satisfactory flotation. This can often be achieved with as little as 5 per cent surface coverage, so that the amount of collector oil used is fairly small. In a typical metal sulphide ore flotation, the slurry will contain approximately 3 tonne of water, 50 g (maximum) of collector oil and 50 g of foaming agent for each tonne of crude ore, and recovery of *c.* 90 per cent of the metal content would normally be achieved.

Sometimes the ore must be pretreated before it will adsorb the additive satisfactorily; for example, zinc sulphide must be pretreated with dilute copper sulphate solution, which deposits copper on the ore surface by electrochemical action. Specificity of flotation may also be achieved by the addition of depressants; for example, cyanide ions prevent ferrous sulphide and zinc sulphide from floating but allow

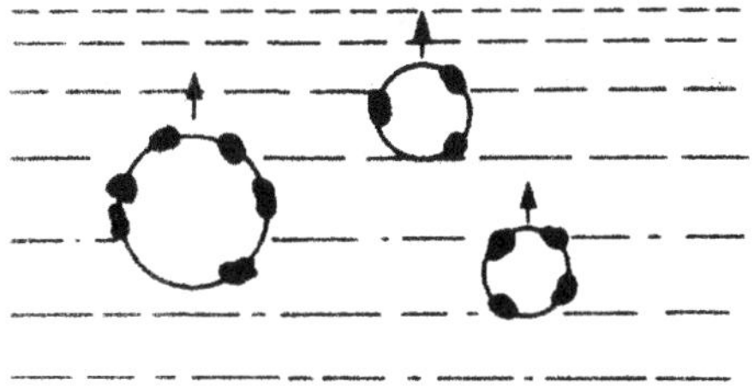

Figure 6.7

lead sulphide to float. In this way the components of a mixed ore can be separated.

The detailed theory of flotation is a little more complicated than indicated by the above account. Bubble adhesion is maximum when there is only 5–15 per cent monolayer coverage by the collector oil and decreases with further coverage. It is thought that when the bubble and particle interfaces merge, penetration of the film of collector oil around the particle by the film of foaming agent around the bubble occurs. This interlocking between the two films stabilises the air bubble–particle system and is, therefore, most favoured when the particles are only partly covered with a film of collector oil. The function of the foaming agent as such may, therefore, be of secondary importance as the particles themselves act as foam stabilisers (see Chapter 10).

Flotation is also used to enrich fuels (e.g. coal and oil) and as a purification procedure for effluents and chemical processing intermediates.

Detergency[82–84]

Detergency is the theory and practice of dirt removal from solid surfaces by surface chemical means. It accounts for the bulk of all surfactant usage

Soaps have been used as detergents for many centuries. Soap normally consists of the sodium or potassium salts of various long-chain fatty acids and is manufactured by the saponification of glyceride oils and fats (e.g. tallow) with NaOH or KOH, giving glycerol as a by-product:

$$\begin{array}{lll} CH_2COOR & CH_2OH & RCOONa \\ | & | & \\ CHCOOR' + 3NaOH = & CHOH & + R'COONa \\ | & | & \\ CH_2COOR'' & CH_2OH & R''COONa \\ \text{fat} & \text{glycerol} & \text{soap} \end{array}$$

The potassium soaps tend to be softer and more soluble in water than the corresponding sodium soaps. Soaps from unsaturated fatty acids are softer than those from saturated fatty acids.

Soap is an excellent detergent but suffers from two main drawbacks: (*a*) it does not function very well in acid solutions because of the formation of insoluble fatty acid, and, (*b*) it forms insoluble precipitates and, hence, a scum with the Ca^{2+} and Mg^{2+} ions in hard water. Additives such as sodium carbonate, phosphates, etc., help to offset these effects. In the last few decades soap has been partly superseded by the use of synthetic (soapless) detergents, which do not suffer to the same extent from these disadvantages. The alkyl sulphates, alkyl-aryl sulphonates and the non-ionic polyethylene oxide derivatives are perhaps the most important. Until recently, the alkyl-aryl sulphonate $(CH_3)_2CH[CH_2CH(CH_3)]_3C_6H_4SO_3^-Na^+$ has found widespread use, but, on account of its non-biodegradability, it has been withdrawn in many countries in favour of the more biodegradable linear isomer and other 'softer' detergents.

Mechanisms of detergency

A satisfactory detergent must possess the following properties:

1. Good wetting characteristics in order that the detergent may come into intimate contact with the surface to be cleaned.
2. Ability to remove or to help remove dirt into the bulk of the liquid.
3. Ability to solubilise or to disperse removed dirt and to prevent it from being redeposited on to the cleaned surface or from forming a scum.

The solid substrate to be cleaned may be a hard surface (e.g. glass, metal, plastics, ceramic), or fibrous (e.g. wool, cotton, synthetic fibres), or a part of the body (skin, hair, teeth). The dirt may be liquid or solid (usually it is a combination of both); it has many possible origins (e.g. skin, food, the atmosphere); it may be polar or non-polar; of small or large particle size; chemically reactive or inert towards the substrate and/or the detergent. In view of the wide variety of possible substrate–dirt systems, the extent to which a general theory of detergent mechanism can be developed is limited. Moreover, when it comes to the formulation of detergents for various types of usage, the situation is even more complex, since performance tends to be judged by criteria which are not wholly related to dirt removal.

Wetting

The wetting of fabrics, as such, is not a critical issue in detergency, since the critical surface tension, γ_c, of fabric surfaces is usually in excess of 40 mN m^{-1} and it is an easy matter to reduce the surface tension of the aqueous bath to below this value. The rate of diffusion of surfactant into porous fabric, however, is important and the choice of surfactant involves a compromise between a small hydrocarbon chain length for rapid diffusion and a longer hydrocarbon chain length for better dirt removal and dispersion characteristics. For alkyl sulphates and alkyl-aryl sulphonates, a chain length of about C_{12} usually gives the best all-round performance in this respect[82].

Dirt removal

The removal of solid dirt can be considered in terms of the surface-energy changes involved. The work of adhesion between a dirt particle and a solid surface (Figure 6.8) is given by

$$W_{SD} = \gamma_{DW} + \gamma_{SW} - \gamma_{SD} \qquad (6.13)$$

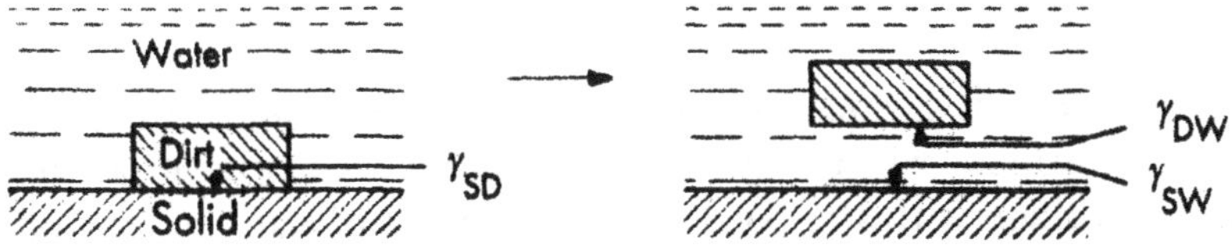

Figure 6.8

The action of the detergent is to lower γ_{DW} and γ_{SW}, thus decreasing W_{SD} and increasing the ease with which the dirt particle can be detached by mechanical agitation.

If the dirt is fluid (oil or grease), its removal can be considered as a contact-angle phenomenon. The addition of detergent lowers the contact angle at the triple solid–oil–water boundary. If $\theta = 0$, the oil will detach spontaneously from the solid substrate. If $0 < \theta < 90°$, the oil can be removed entirely by mechanical means (Figure 6.9a); but if $90° < \theta < 180°$, only part of the oil can be detached by mechanical means and some will remain attached to the solid substrate (Figure 6.9b). A different mechanism, (e.g. solubilisation) is required to remove this residual oil. Relating to this roll-up mechanism, increasing the temperature has a marked effect on detergent efficiency up to about 45°C (most fats melt below this temperature) and little effect between about 45°C and just below the boiling point.

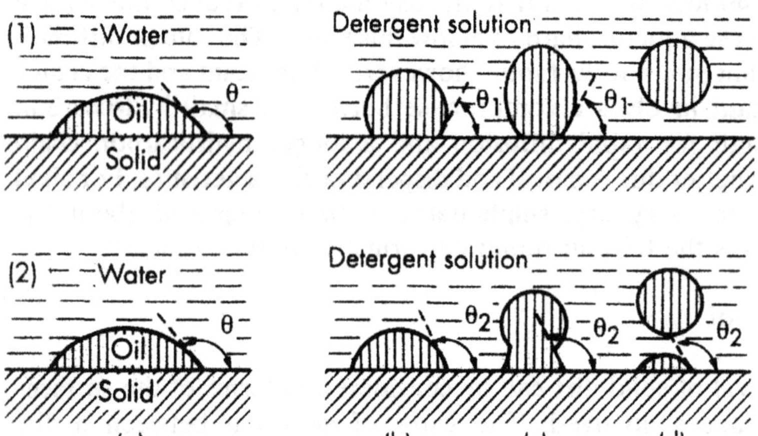

Figure 6.9 Detachment of oily dirt from a solid surface. The sequences (left to right) show: (a) the substrate/dirt system in contact with pure water, (b) the lowering of contact angle caused by detergent [(1) $\theta < 90°$, (2) $\theta > 90°$], and (c) and (d) mechanical (hydraulic) detachment of oil droplets

It can be seen that surfactants which adsorb at the solid–water and dirt–water interfaces will be the best detergents. Adsorption at the air–water interface with the consequent lowering of surface tension and foaming is not necessarily an indication of detergent effectiveness; for example, non-ionic detergents usually have excellent detergent action yet are poor foaming agents, and the psychological tendency to correlate these two properties has somewhat restricted their acceptance for household usage.

Redeposition of dirt

Redeposition of dirt can be prevented by the charge and hydration barriers which are set up as a result of detergent molecules being adsorbed on to the cleaned material and on to the dirt particles. Since the substrate and dirt surfaces tend to be negatively charged (see page 175), anionic detergents tend to be more effective than cationic detergents. Non-ionic detergents are also effective in this respect as a result of strong hydration of the poly(ethylene oxide) chains. Mixed anionic plus non-ionic detergents usually out-perform anionics alone.

The most successful detergents are those forming micelles, and this originally led to the opinion that micelles are directly involved in detergent action, their role probably being that of solubilising oily material. However, detergent action is dependent upon the concentration of unassociated surfactant and practically unaffected by the presence of micelles (other than as a reservoir for replenishing the unassociated surfactant adsorbed from solution). It appears, therefore, that the molecular properties of surfactants associated with good detergent action also lead to micelle formation as a competing rather than as a contributing process[2].

Detergent additives

It is general practice to incorporate 'builders', such as silicates, pyrophosphates and tripolyphosphates, which are not surface-active themselves but which improve the performance of the detergent. Builders fulfil a number of functions, the most important being to sequester (form soluble non-adsorbed complexes with) Ca^{2+} and Mg^{2+} ions and act as deflocculating agents, thus helping to avoid scum formation and dirt redeposition. The builders also help by producing the mildly alkaline conditions which are favourable to detergent action. Alternatives to phosphates are being sought for environmental reasons.

Sodium carboxymethyl cellulose improves detergent performance in washing textile fabrics, particularly cotton, by forming a protective hydrated adsorbed layer on the cleaned fabric which helps to prevent

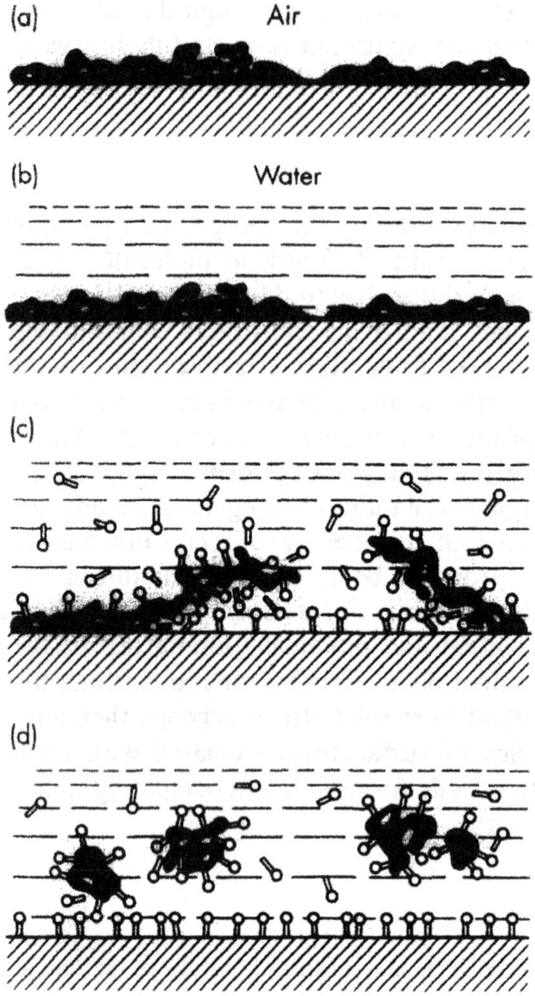

Figure 6.10 Removal of dirt from a solid surface by detergent and mechanical action[85]. (a) Surface covered with greasy dirt. (b) Water by itself fails to dislodge the dirt mainly because of its high surface tension and inefficient wetting action. (c) Detergent is added to the water. The hydrophobic parts of the detergent molecules line up both on the dirt and on the solid surface, thus reducing the adhesion of the dirt to the solid. The dirt may now be dislodged by mechanical action. (d) Dirt is held suspended in the solution because the detergent molecules form an adsorbed layer on the cleaned surface and around the dirt particles (*By courtesy of The Scientific American Inc.*)

dirt redeposition. Optical brighteners are commonly incorporated into detergents used for washing textile fabrics. These are fluorescent dyes which absorb ultraviolet light and emit blue light which masks any yellow tint which may develop in white fabrics.

Adsorption from solution[86]

To conclude this chapter, some general comments concerning the adsorption of material from solutions on to solid surfaces are appropriate. Adsorption from solution is important in many practical situations, such as those in which modification of the solid surface is of primary concern (e.g. the use of lyophilic material to stabilise dispersions; see page 235) and those which involve the removal of unwanted material from the solution (e.g. the clarification of sugar solutions with activated charcoal). The adsorption of ions from electrolyte solutions and a special case of ion adsorption, that of ion exchange, are discussed in Chapter 7. Adsorption processes are, of course, most important in *chromatography*; however, an account of chromatography is not included in this book (*a*) because other processes, such as partition and/or molecular sieving, may also be involved to a greater or lesser extent, depending on the type of chromatographic separation being considered and (*b*) because chromatography is far too extensive a subject to permit adequate treatment in a relatively small space.

Solution adsorption isotherms

Experimentally, the investigation of adsorption from solution is much simpler than that of gas adsorption. A known mass of adsorbent solid is shaken with a known volume of solution at a given temperature until there is no further change in the concentration of the supernatant solution. This concentration can be determined by a variety of methods involving chemical or radiochemical analysis, colorimetry, refractive index, etc. The experimental data are usually expressed in terms of an *apparent adsorption isotherm* in which the amount of solute adsorbed at a given temperature per unit mass of adsorbent – as calculated from the decrease (or increase) of solution concentration – is plotted against the equilibrium concentration.

The theoretical treatment of adsorption from solution, however, is, in general, more complicated than that of gas adsorption, since adsorption from solution always involves competition between solute(s) and solvent or between the components of a liquid mixture for the adsorption sites. Consider, for example, a binary liquid mixture in contact with a solid. Zero adsorption refers to uniform mixture composition right up to the solid surface, even though (unlike zero gas adsorption) both components are, in fact, present at the solid surface. If the proportion of one of the components at the surface is greater than its proportion in bulk, then that component is positively adsorbed and, consequently, the other component is negatively adsorbed. Apparent, rather than true, adsorption isotherms are, therefore, calculated from changes in solution concentration. Examples of apparent adsorption isotherms for binary liquid mixtures are given in Figure 6.11. Within the context of certain assumptions,

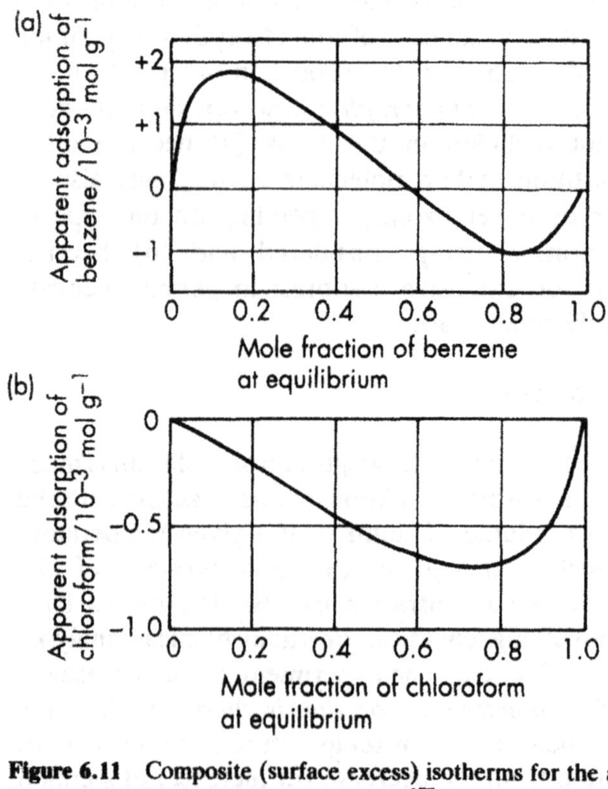

Figure 6.11 Composite (surface excess) isotherms for the adsorption of (a) benzene from solution in methanol on to charcoal[177] and (b) chloroform from solution in carbon tetrachloride on to charcoal[178] (*By courtesy of (a) American Chemical Society, (b) Journal of the Chemical Society*)

the individual adsorption isotherms can be calculated from the apparent (or composite) adsorption isotherm together with appropriate vapour adsorption data[63].

Adsorption from solution behaviour can often be predicted qualitatively in terms of the polar/non-polar nature of the solid and of the solution components. This is illustrated by the isotherms shown in Figure 6.12 for the adsorption of fatty acids from toluene solution on to silica gel and from aqueous solution on to carbon.

A polar adsorbent will tend to adsorb polar adsorbates strongly and non-polar adsorbates weakly, and vice versa. In addition, polar solutes will tend to be adsorbed strongly from non-polar solvents (low solubility) and weakly from polar solvents (high solubility), and vice versa. For the isotherms represented in Figure 6.12a the solid is polar, the solutes are amphiphilic and the solvent is non-polar. Fatty acid adsorption is, therefore, strong compared with that of the solvent. In accord with the above generalisations, the amount of fatty acid adsorbed at a given concentration decreases with increasing length of non-polar hydrocarbon chain – i.e. acetic > propionic > butyric. In Figure 6.12b the solid is non-polar and the solvent is polar, so, again, fatty acid adsorption is strong compared with that of the solvent. However, since the adsorbent is non-polar and the solvent polar, the amount of fatty acid adsorbed at a given concentration now increases with increasing length of non-polar hydrocarbon chain – i.e. butyric > propionic > acetic.

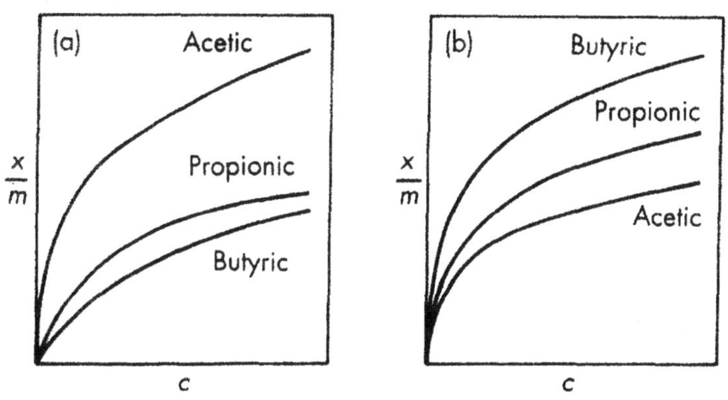

Figure 6.12 Adsorption isotherms for fatty acids: (a) from toluene solutions on to silica gel and (b) from aqueous solutions on to charcoal

Isotherm equations, surface areas

In adsorption from solution, physical adsorption is far more common than chemisorption. However, chemisorption is sometimes possible; for example, fatty acids are chemisorbed from benzene solutions on nickel and platinum catalysts.

Solute adsorption is usually restricted to a monomolecular layer, since the solid–solute interactions, although strong enough to compete successfully with the solid–solvent interactions in the first adsorbed monolayer, do not do so in subsequent monolayers. Multilayer adsorption has, however, been observed in a number of cases, being evident from the shape of the adsorption isotherms and from the impossibly small areas per adsorbed molecule calculated on the basis of monomolecular adsorption.

The adsorption from solution of polymers has been studied extensively. The amount of polymer adsorbed usually reaches a limiting value as the concentration of polymer in solution is increased, but this value is usually well in excess of that which would be expected for a monomolecular layer of polymer adsorbed flat on the solid surface. This suggests that the adsorbed polymer is anchored to the surface only at a few points, with the remainder of the polymer in the form of loops and ends moving more or less freely in the liquid phase[179].

The Langmuir and Freundlich equations (see page 128) are frequently applied to adsorption from solution data, for which they take the form

$$x/m = \frac{(x/m)_{\text{max.}}\,ac}{1+ac} \tag{6.14}$$

and

$$x/m = kc^{1/n} \tag{6.15}$$

respectively, where x is the amount of solute adsorbed by a mass m of solid, c is the equilibrium solution concentration, and a, k and n are constants.

If the monolayer capacity $(x/m)_{\text{max.}}$ can be estimated (either directly from the actual isotherm or indirectly by applying the Langmuir equation) and if the effective area occupied by each adsorbed molecule is known, the specific surface area of the solid can be calculated as described on page 134–36 for gas adsorption.

Adsorption from solution has the merit of being experimentally less demanding than gas adsorption; however, the problems in interpretation are far greater.

Since relatively large and asymmetric adsorbate molecules, such as moderately long-chain fatty acids and various dyestuffs, are usually involved in adsorption from solution, it is necessary to make assumptions regarding their orientation and packing efficiency in calculating their effective surface coverage. In view of the uncertainties involved in such calculations, it is usually desirable to calibrate a particular adsorption from solution system with the aid of a surface area determined by a less complex method, such as nitrogen adsorption. Adsorption from solution can then provide a convenient technique for determining specific surface areas.

7 Charged interfaces

The electric double layer

Most substances acquire a surface electric charge when brought into contact with a polar (e.g. aqueous) medium, possible charging mechanisms (elaborated below) being ionisation, ion adsorption and ion dissolution. This surface charge influences the distribution of nearby ions in the polar medium. Ions of opposite charge (*counter-ions*) are attracted towards the surface and (less important) ions of like charge (*co-ions*) are repelled away from the surface. This, together with the mixing tendency of thermal motion, leads to the formation of an electric double layer made up of the charged surface and a neutralising excess of counter-ions over co-ions distributed in a diffuse manner in the polar medium. The theory of the electric double layer deals with this distribution of ions and, hence, with the magnitude of the electric potentials which occur in the locality of the charged surface. This is a necessary first step towards understanding many of the experimental observations concerning the electrokinetic properties, stability, etc., of charged colloidal systems.

Origin of the charge at surfaces

Ionisation

Proteins acquire their charge mainly through the ionisation of carboxyl and amino groups to give COO^- and NH_3^+ ions. The ionisation of these groups, and so the net molecular charge, depends strongly on the pH of the solution. At low pH a protein molecule will be positively charged and at high pH it will be negatively charged. The pH at which the net charge (and electrophoretic mobility) is zero is called the *iso-electric point* (see Table 2.3 and Figure 7.7).

Ion adsorption

A net surface charge can be acquired by the unequal adsorption of oppositely charged ions. Ion adsorption may involve positive or negative surface excess concentrations.

Surfaces in contact with aqueous media are more often negatively charged than positively charged. This is a consequence of the fact that cations are usually more hydrated than anions and so have the greater tendency to reside in the bulk aqueous medium; whereas the smaller, less hydrated and more polarising anions have the greater tendency to be specifically adsorbed.

Hydrocarbon oil droplets and even air bubbles suspended in water and in most aqueous electrolyte solutions have negative electrophoretic mobilities (i.e. they migrate towards the anode under the influence of an applied electric field)[180]. This net negative charge is explained in terms of negative adsorption of ions. The addition of simple electrolytes, such as NaCl results in an increase in the surface tension of water (see Figure 4.11) and in the interfacial tension between hydrocarbon oil and water. This is interpreted via the Gibbs equation (page 80–2) in terms of a negative surface excess ionic concentration. The surface excess concentrations of hydrogen and hydroxyl ions will also be negative. Presumably, cations move away from the air bubble–water and oil–water interfaces more than anions, leaving the kinetic units (which will include some aqueous medium close to the interfaces) with net negative charges.

Preferential negative adsorption of hydrogen ions compared with hydroxyl ions is reflected in the electrophoretic mobility–pH curve for hydrocarbon oil droplets (see Figure 7.7). The magnitude of the electrophoretic mobilities of inert particles such as hydrocarbon oil droplets (*c* 0 to -6×10^{-8} m^2 s^{-1} V^{-1}) is comparable with those of simple ions (e.g. -7.8×10^{-8} m^2 s^{-1} V^{-1} for Cl^- ions at infinite dilution in aqueous solution at 25°C), which, in view of their relatively large size, reflects a high charge number.

Surfaces which are already charged (e.g. by ionisation) usually show a preferential tendency to adsorb counter-ions, especially those with a high charge number. It is possible for counter-ion adsorption to cause a reversal of charge.

If surfactant ions are present, their adsorption will usually determine the surface charge.

Hydrated (e.g. protein and polysaccharide) surfaces adsorb ions less readily than hydrophobic (e.g. lipid) surfaces.

Ion dissolution

Ionic substances can acquire a surface charge by virtue of unequal dissolution of the oppositely charged ions of which they are composed.

Silver iodide particles in aqueous suspension are in equilibrium with a saturated solution of which the solubility product, $a_{Ag^+}a_{I^-}$, is about 10^{-16} at room temperature. With excess I^- ions, the silver iodide particles are negatively charged; and with sufficient excess Ag^+ ions, they are positively charged. The zero point of charge is not at pAg 8 but is displaced to pAg 5.5 (pI 10.5), because the smaller and more mobile Ag^+ ions are held less strongly than the I^- ions in the silver iodide crystal lattice. The silver and iodide ions are referred to as *potential-determining* ions, since their concentrations determine the electric potential at the particle surface. Silver iodide sols have been used extensively for testing electric double layer and colloid stability theories.

In a similar way, hydrogen and hydroxyl ions are potential-determining for hydrous metal oxide sols:

$$—M—OH + H^+ = —M—OH_2^+$$

$$—M—OH + OH^- = —M—O^- + H_2O$$

The surface electrochemistry of hydrous metal oxides is actually much more complicated than this, with a range of individual inorganic reactions possible, depending on factors such as surface crystal structure.

Adsorption and orientation of dipoles

Adsorption of dipolar molecules will not contribute to a net surface charge, but the presence of a layer of orientated dipolar molecules at the surface may make a significant contribution to the nature of the electric double layer.

The diffuse double layer

The electric double layer can be regarded as consisting of two regions: an inner region which may include adsorbed ions, and a diffuse region in which ions are distributed according to the influence of electrical forces and random thermal motion. The diffuse part of the double layer will be considered first.

Quantitative treatment of the electric double layer represents an extremely difficult and in some respects unresolved problem. The requirement of overall electroneutrality dictates that, for any dividing surface, if the charge per unit area is $+\sigma$ on one side of the surface, it must be $-\sigma$ on the other side. It follows, therefore, that the magnitude of σ will depend on the location of the surface. Surface location is not a straightforward matter owing to the geometric and chemical heterogeneity which generally exists. It follows, furthermore, that electric double-layer parameters (potentials, surface charge densities, distances) are not amenable to unequivocal definition. Despite this, however, various simplifications and approximations can be made which allow double-layer theory to be developed to a high level of sophistication and usefulness.

The simplest quantitative treatment of the diffuse part of the double layer is that due to Gouy (1910) and Chapman (1913), which is based on the following model:

1. The surface is assumed to be flat, of infinite extent and uniformly charged.
2. The ions in the diffuse part of the double layer are assumed to be point charges distributed according to the Boltzmann distribution.
3. The solvent is assumed to influence the double layer only through its dielectric constant, which is assumed to have the same value throughout the diffuse part.
4. A single symmetrical electrolyte of charge number z will be assumed. This assumption facilitates the derivation while losing little owing to the relative unimportance of co-ion charge number.

Let the electric potential be ψ_0 at a flat surface and ψ at a distance x from the surface in the electrolyte solution. Taking the surface to be positively charged (Figure 7.1) and applying the Boltzmann distribution,

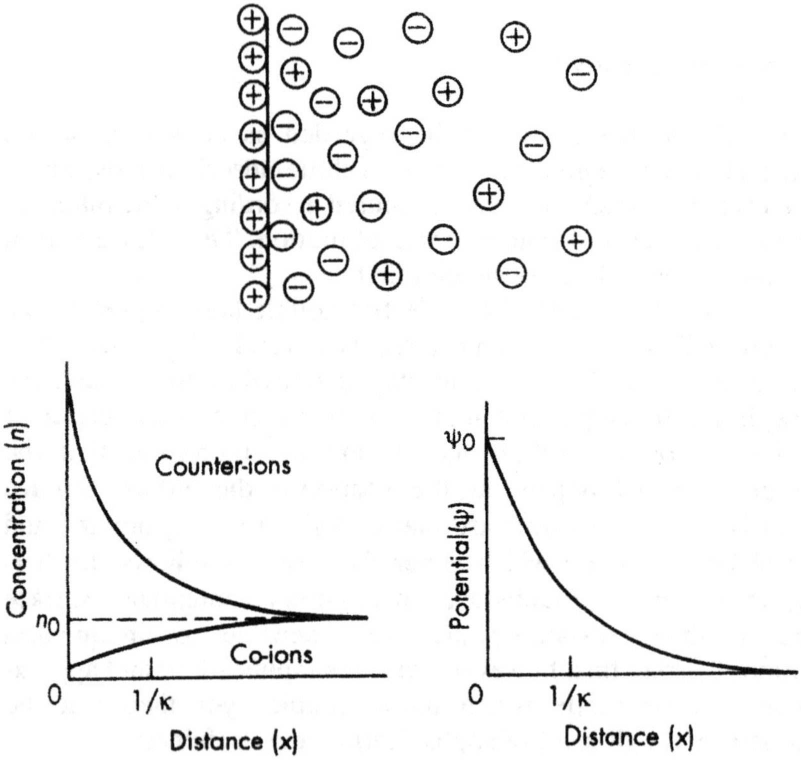

Figure 7.1 Schematic representation of a diffuse electric double layer

$$n_+ = n_0 \exp\left[\frac{-ze\psi}{kT}\right] \qquad \text{and} \qquad n_- = n_0 \exp\left[\frac{+ze\psi}{kT}\right]$$

where n_+ and n_- are the respective numbers of positive and negative ions per unit volume at points where the potential is ψ (i.e. where the electric potential energy is $ze\psi$ and $-ze\psi$, respectively), and n_0 is the corresponding bulk concentration of each ionic species.

The net volume charge density ρ at points where the potential is ψ is, therefore, given by

$$\begin{aligned} \rho &= ze(n_+ - n_-) \\ &= zen_0\left(\exp\left[\frac{-ze\psi}{kT}\right] - \exp\left[\frac{+ze\psi}{kT}\right]\right) \\ &= -2zen_0 \sinh\frac{ze\psi}{kT} \end{aligned} \tag{7.1}$$

ρ is related to ψ by Poisson's equation, which for a flat double layer takes the form

$$\frac{d^2\psi}{dx^2} = -\frac{\rho}{\epsilon} \tag{7.2}$$

where ϵ is the permittivity*.

Combination of equations (7.1) and (7.2) gives

$$\frac{d^2\psi}{dx^2} = \frac{2zen_0}{\epsilon}\sinh\frac{ze\psi}{kT} \tag{7.3}$$

The solution[87] of this expression, with the boundary conditions ($\psi = \psi_0$ when $x = 0$; and $\psi = 0$, $d\psi/dx = 0$ when $x = \infty$) taken into account, can be written in the form

$$\psi = \frac{2kT}{ze}\ln\left(\frac{1+\gamma\exp[-\kappa x]}{1-\gamma\exp[-\kappa x]}\right) \tag{7.4}$$

where

$$\gamma = \frac{\exp[ze\psi_0/2kT]-1}{\exp[ze\psi_0/2kT]+1} \tag{7.5}$$

and

$$\kappa = \left(\frac{2e^2n_0z^2}{\epsilon kT}\right)^{1/2} = \left(\frac{2e^2N_Acz^2}{\epsilon kT}\right)^{1/2} = \left(\frac{2F^2cz^2}{\epsilon RT}\right)^{1/2} \tag{7.6}$$

where N_A is Avogadro's constant and c is the concentration of electrolyte.

If $ze\psi_0/2kT \ll 1$ (kT/e = 25.6 mV at 25°C), the Debye–Hückel approximation,

$$\left(\exp\left[\frac{ze\psi_0}{2kT}\right] \approx 1 + \frac{ze\psi_0}{2kT}\right)$$

can be made and equations (7.4) and (7.5) simplify to

*The permittivity of a material is the constant ϵ in the rationalised expression, $F = (Q_1Q_2)/(4\pi\epsilon r^2)$, where F is the force between charges Q_1 and Q_2 separated by a distance r. The permittivity of a vacuum ϵ_0 according to this definition is equal to 8.854 × 10^{-12} kg^{-1} m^{-3} s^4 A^2. The dielectric constant of a material is equal to the ratio between its permittivity and the permittivity of a vacuum, and is a dimensionless quantity.

$$\psi = \psi_0 \exp[-\kappa x] \tag{7.7}$$

which shows that at low potentials the potential decreases exponentially with distance from the charged surface. Close to the charged surface, where the potential is likely to be relatively high and the Debye–Hückel approximation inapplicable, the potential is predicted to decrease at a greater than exponential rate.

The potential ψ_0 can be related to the charge density σ_0 at the surface by equating the surface charge with the net space charge in the diffuse part of the double layer $\left(\text{i.e. } \sigma_0 = -\int_0^\infty \rho \mathrm{d}x\right)$ and applying the Poisson–Boltzmann distribution. The resulting expression is

$$\sigma_0 = (8n_0\epsilon kT)^{1/2} \sinh \frac{ze\psi_0}{2kT} \tag{7.8}$$

which at low potentials reduces to

$$\sigma_0 = \epsilon\kappa\psi_0 \tag{7.9}$$

The surface potential ψ_0, therefore, depends on both the surface charge density σ_0 and (through κ) on the ionic composition of the medium. If the double layer is compressed (i.e. κ increased), then either σ_0 must increase, or ψ_0 must decrease, or both.

In many colloidal systems, the double layer is created by the adsorption of potential-determining ions; for example, the potential ψ_0 at the surface of a silver iodide particle depends on the concentration of silver (and iodide) ions in solution. Addition of inert electrolyte increases κ and results in a corresponding increase of surface charge density caused by the adsorption of sufficient potential-determining silver (or iodide) ions to keep ψ_0 approximately constant. In contrast, however, the charge density at an ionogenic surface remains constant on addition of inert electrolyte (provided that the extent of ionisation is unaffected) and ψ_0 decreases.

From equation (7.9) it can be seen that, at low potentials, a diffuse double layer has the same capacity as a parallel plate condenser with a distance $1/\kappa$ between the plates. It is customary to refer to $1/\kappa$ (the distance over which the potential decreases by an exponential factor at low potentials) as the 'thickness' of the diffuse double layer.

For an aqueous solution of a symmetrical electrolyte at 25°C, equation (7.6) becomes

$$\kappa = 0.329 \times 10^{10} \left(\frac{cz^2}{\text{mol dm}^{-3}} \right)^{\frac{1}{2}} \text{m}^{-1} \tag{7.10}$$

For a 1–1 electrolyte the double layer thickness is, therefore, about 1 nm for a 10^{-1} mol dm^{-3} solution and about 10 nm for a 10^{-3} mol dm^{-3} solution. For unsymmetrical electrolytes the double layer thickness can be calculated by taking z to be the counter-ion charge number.

The Poisson–Boltzmann distribution for a spherical interface takes the form

$$\nabla^2 \psi = \frac{1}{r^2} \frac{\text{d}}{\text{d}r} \left(\frac{r^2 \text{d}\psi}{\text{d}r} \right) = \frac{2zen_0}{\epsilon} \sinh \frac{ze\psi}{kT} \tag{7.11}$$

where r is the distance from the centre of the sphere. This expression cannot be integrated analytically without approximation to the exponential terms. If the Debye–Hückel approximation is made, the equation reduces to

$$\nabla^2 \psi = \kappa^2 \psi \tag{7.12}$$

which, on integration (with the boundary conditions, $\psi = \psi_0$ at $r = a$ and $\psi = 0$, $\text{d}\psi/\text{d}r = 0$ at $r = \infty$, taken into account) gives

$$\psi = \psi_0 \frac{a}{r} \exp[-\kappa)r - a)] \tag{7.13}$$

Unfortunately, the Debye–Hückel approximation ($z\psi \ll c.$ 25 mV) is often not a good one in the treatment of colloid and surface phenomena. Unapproximated, numerical solutions of equation (7.11) have been computed.[88]

The inner part of the double layer

The treatment of the diffuse double layer outlined in the last section is based on an assumption of point charges in the electrolyte medium. The finite size of the ions will, however, limit the inner boundary of the diffuse part of the double layer, since the centre of an ion can only approach the surface to within its hydrated radius without becoming specifically adsorbed. Stern (1924) proposed a model in which the double layer is divided into two parts separated by a plane (the Stern plane) located at about a hydrated ion radius from the surface, and also considered the possibility of specific ion adsorption.

Specifically adsorbed ions are those which are attached (albeit temporarily) to the surface by electrostatic and/or van der Waals forces strongly enough to overcome thermal agitation. They may be dehydrated, at least in the direction of the surface. The centres of any specifically adsorbed ions are located in the Stern layer – i.e. between the surface and the Stern plane. Ions with centres located beyond the Stern plane form the diffuse part of the double layer, for which the Gouy–Chapman treatment outlined in the previous section, with ψ_0 replaced by ψ_d, is considered to be applicable.

The potential changes from ψ_0 (the surface or wall potential) to ψ_d (the Stern potential) in the Stern layer, and decays from ψ_d to zero in the diffuse double layer.

In the absence of specific ion adsorption, the charge densities at the surface and at the Stern plane are equal and the capacities of the Stern layer (C_1) and of the diffuse layer (C_2) are given by

$$C_1 = \frac{\sigma_0}{\psi_0 - \psi_d} \quad \text{and} \quad C_2 = \frac{\sigma_0}{\psi_d}$$

from which

$$\psi_d = \frac{C_1 \psi_0}{C_1 + C_2} \tag{7.14}$$

When specific adsorption takes place, counter-ion adsorption usually predominates over co-ion adsorption and a typical double layer situation would be that depicted in Figure 7.2. It is possible, especially with polyvalent or surface-active counter-ions, for reversal of charge to take place within the Stern layer – i.e. for ψ_0 and ψ_d to have opposite signs (Figure 7.3a). Adsorption of surface-active co-ions could create a situation in which ψ_d has the same sign as ψ_0 and is greater in magnitude (Figure 7.3b).

Stern assumed that a Langmuir-type adsorption isotherm could be used to describe the equilibrium between ions adsorbed in the Stern layer and those in the diffuse part of the double layer. Considering only the adsorption of counter-ions, the surface charge density σ_1 of the Stern layer is given by the expression

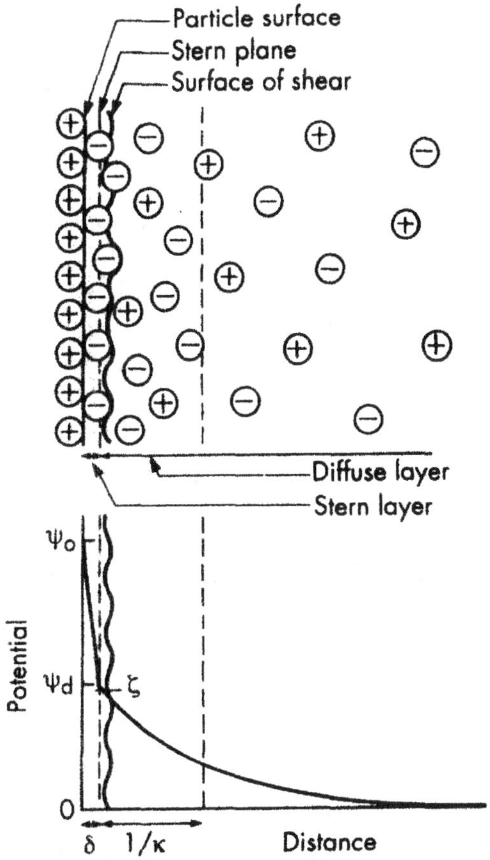

Figure 7.2 Schematic representation of the structure of the electric double layer according to Stern's theory

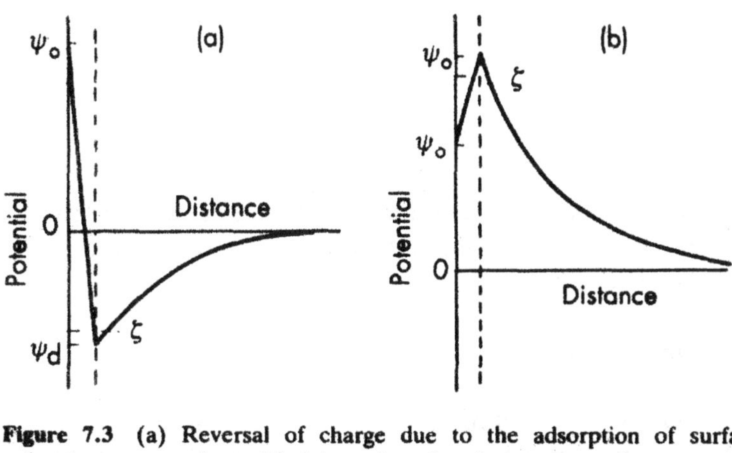

Figure 7.3 (a) Reversal of charge due to the adsorption of surface-active or polyvalent counter-ions. (b) Adsorption of surface-active co-ions

$$\sigma_1 = \frac{\sigma_m}{1 + \frac{N_A}{n_0 v_m} \exp\left[\frac{ze\psi_d + \phi}{kT}\right]} \tag{7.15}$$

where σ_m is the surface charge density corresponding to a monolayer of counter-ions, N_A is Avogadro's constant and V_m is the molar volume of the solvent. The adsorption energy is divided between electrical ($ze\phi_d$) and van der Waals (ϕ) terms.

Treating the Stern layer as a molecular condenser of thickness δ and with a permittivity ϵ',

$$\sigma_0 = \frac{\epsilon'}{\delta}(\psi_0 - \psi_d) \tag{7.16}$$

where σ_0 is the charge density at the particle surface.

For overall electrical neutrality throughout the whole of the double layer,

$$\sigma_0 + \sigma_1 + \sigma_2 = 0 \tag{7.17}$$

where σ_2 is the surface charge density of the diffuse part of the double layer and is given by equation (7.8) with the sign reversed and with ψ_0 replaced by ψ_d.

Substituting from equations (7.16), (7.15) and (7.8) into equation (7.17) gives a complete expression for the Stern model of the double layer:

$$\frac{\epsilon'}{\delta}(\psi_0 - \psi_d) + \frac{\sigma_m}{1 + \frac{N_A}{n_0 V_m} \exp\left[\frac{ze\psi_d + \phi}{kT}\right]} - (8n_0\epsilon kT)^{1/2} \sinh\frac{ze\psi_d}{2kT} = 0 \tag{7.18}$$

This expression contains a number of unknown quantities; however, as indicated below, some information can be derived about these from other sources.

Permittivity of the Stern layer

The total capacity C of the double layer has been determined from electrocapillary measurements for mercury–aqueous electrolyte interfaces[89], and from potentiometric titration measurements for silver iodide–aqueous electrolyte interfaces[181]. If the double layer is treated as two capacitors in series, then

$$\frac{1}{C} = \frac{1}{C_1} + \frac{1}{C_2}$$

The capacity C_2 of the diffuse part of the double layer can be calculated. At low potentials (*see* equation (7.9)):

$$C_2 = \frac{\sigma_2}{\psi_d} = \epsilon\kappa$$

$$= 2.28\left(\frac{cz^2}{\text{mol dm}^{-3}}\right)^{1/2} \text{F m}^{-2} \text{ for aqueous electrolyte at 25°C}$$

The capacity of the Stern layer ($C_1 = \epsilon'/\delta$) does not depend on electrolyte concentration except in so far as ϵ' is affected. In the case of the silver iodide–aqueous electrolyte interface, Stern layer capacities of *c.* 0.1 to 0.2 F m^{-2} have been calculated; taking $\delta = 5 \times 10^{-10}$ m, this corresponds to a dielectric constant in the Stern layer of *c.* 5–10, which, compared with the normal value of *c.* 80 for water, suggests considerable ordering of water molecules close to the surface.

Stern potentials and electrokinetic (zeta) potentials

ψ_d can be estimated from electrokinetic measurements. Electrokinetic behaviour (discussed in the following sections of this chapter) depends on the potential at the surface of shear between the charged surface and the electrolyte solution. This potential is called the *electrokinetic* or ζ (*zeta*) potential. The exact location of the shear plane (which, in reality, is a region of rapidly changing viscosity) is another unknown feature of the electric double layer. In addition to ions in the Stern layer, a certain amount of solvent will probably be bound to the charged surface and form a part of the electrokinetic unit. It is, therefore, reasonable to suppose that the shear plane is usually located at a small distance further out from the surface than the Stern plane and that ζ is, in general, marginally smaller in magnitude than ψ_d (see Figures 7.2 and 7.3). In tests of double-layer theory it is customary to assume identity of ψ_d and ζ, and the bulk of experimental evidence suggests that errors introduced through this assumption are usually small, especially at lyophobic surfaces. Any difference between ψ_d and ζ will clearly be most pronounced at high potentials ($\zeta = 0$ when $\psi_d = 0$), and at high electrolyte concentration (compression of the diffuse part of the double layer will cause more of the potential drop from ψ_d to zero to take place within the shear plane). The adsorption of non-ionic surfactant would result in the surface of shear being located at a relatively large distance from the Stern plane and a zeta potential significantly lower than ψ_d.

Surface potentials

For an interface such as silver iodide–electrolyte solution the electric potential difference between the solid interior and the bulk solution varies according to the Nernst equation:

$$\frac{d\phi}{d(pAg)} = \frac{-2.303\ RT}{F} \quad (= -59\ \text{mV at } 25°\text{C})$$

ϕ is made up of two terms, ψ_0 and χ. Changes in the χ (*chi*) potential arise from the adsorption and/or orientation of dipolar (e.g. solvent) molecules at the surface or from the displacement of oriented dipolar molecules from the surface. Such effects are difficult to estimate. It is often assumed that χ remains constant during variations of the surface potential and that an expression such as

$$\frac{d\psi_0}{d(pAg)} = \frac{-2.303\ RT}{F} \tag{7.19}$$

is justified. It is also assumed in this expression that no double layer occurs within the solid; this may not be so, since the excess Ag^+ (or I^-) ions of the silver iodide particle do not necessarily all reside at the particle surface. The zero point of zeta potential (which is a measurable quantity, pAg 5.5 at 25°C for AgI in aqueous dispersion medium) can be identified with $\psi_0 = 0$ if specific adsorption of non-potential-determining ions is assumed to be absent. ψ_0 can, therefore, be calculated for a given pAg on the basis of the assumptions outlined here.

Experimentally $\left(\frac{d\zeta}{d(pAg)}\right)_{\zeta \to 0}$ is found to be about -40 mV at 25°C for silver iodide[90] (Figure 7.4) and silver bromide[182] hydrosols prepared by simple mixing, and not -59 mV. Assuming that ζ can be

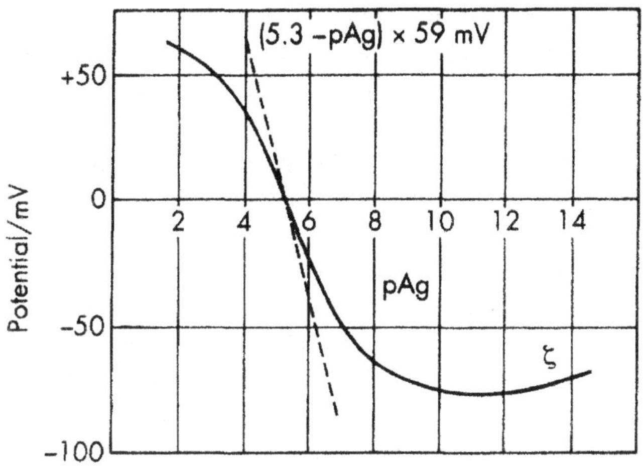

Figure 7.4 Zeta potentials for silver iodide sols prepared by simple mixing[90]. ζ calculated from Smoluchowski equation (*By courtesy of Elsevier Publishing Company*)

identified with ψ_d and making the assumptions outlined in the last paragraph,

$$\left(\frac{d\zeta}{d(pAg)}\right)_{\zeta\to 0} = \left(\frac{d\psi_0}{d(pAg)}\right)\left(\frac{d\psi_d}{d\psi_0}\right)_{\zeta\to 0}$$

$$= -59\frac{C_1}{C_1+C_2}\,\text{mV} \tag{7.20}$$

Values of $\left(\frac{d\zeta}{d(pAg)}\right)_{\zeta\to 0}$ calculated from double-layer capacities using the above expression seem to be, at least semiquantitatively, in accord with the experimental values (from electrophoretic measurements)[91], which suggests that the assumptions involved may be valid. However, this is a topic which has not yet been subjected to sufficient systematic investigation for firm conclusions to be drawn.

From the discussion so far it can be appreciated that the Stern model of the electric double layer presents only a rough picture of what is undoubtedly a most complex situation. Nevertheless, it provides a good basis for interpretating, at least semiquantitatively, most experimental observations connected with electric double layer phenomena. In particular, it helps to account for the magnitude of electrokinetic potentials (rarely in excess of 75 mV) compared with thermodynamic potentials (which can be several hundred millivolts).

A refinement of the Stern model has been proposed by Grahame[89], who distinguishes between an 'outer Helmholtz plane' to indicate the closest distance of approach of hydrated ions (i.e. the same as the Stern plane) and an 'inner Helmholtz plane' to indicate the centres of ions, particularly anions, which are dehydrated (at least in the direction of the surface) on adsorption.

Finally, both the Gouy–Chapman and the Stern treatments of the double layer assume a uniformly charged surface. The surface charge, however, is not 'smeared out' but is located at discrete sites on the surface. When an ion is adsorbed into the inner Helmholtz plane, it will rearrange neighbouring surface charges and, in doing so, impose a self-atmosphere potential ϕ_β on itself (a two-dimensional analogue of the self-atmosphere potential occurring in the Debye–Hückel theory of strong electrolytes). This 'discreteness of charge' effect can be incorporated into the Stern–Langmuir expression, which now becomes

$$\sigma_1 = \frac{\sigma_m}{1 + \dfrac{N_A}{n_0 V_m} \exp\left[\dfrac{ze(\psi_d + \phi_\beta) + \phi}{kT}\right]} \tag{7.21}$$

The main consequence of including this self-atmosphere term is that the theory now predicts that, under suitable conditions, ψ_d goes through a maximum as ψ_0 is increased. The discreteness of charge effect, therefore, explains, at least qualitatively, the experimental observations that both zeta potentials (see Figure 7.4) and coagulation concentrations (see Chapter 8) for sols such as silver halides go through a maximum as the surface potential is increased[183].

Ion exchange

Ion exchange involves an electric double layer situation in which two kinds of counter-ions are present, and can be represented by the equation

$$\text{RA} + \text{B} = \text{RB} + \text{A}$$

where R is a charged porous solid. Counter-ions A and B compete for position in the electric double layer around R, and, in this respect, concentration and charge number are of primary importance. R may be a cation exchanger (fixed negatively charged groups, such as $-SO_3^-$ or $-COO^-$) or an anion exchanger (fixed positively charged groups, such as $-NH_3^+$). A range of highly porous synthetic cation and anion exchange resins are available commercially. The porosity of the resin facilitates fairly rapid ion exchange.

The most important applications of ion exchange are the softening of water and the 'deionisation' of water.

In the first of these processes, hard water is passed through a column of a cation exchange resin usually saturated with sodium counter-ions. The doubly charged (and, therefore, more strongly adsorbed) calcium ions in the water exchange with the singly charged sodium ions in the resin, thus softening the water. Regeneration of the resin is effected by passing a strong solution of sodium chloride through the column.

The 'deionisation' of water involves both anion and cation exchange. A cation exchange resin saturated with hydrogen ions and an anion exchange resin saturated with hydroxyl ions are used, often in the form of a mixed ion exchange resin. These hydrogen and hydroxyl ions exchange with the cations and anions in the water sample and combine to form water.

Ion exchange has many preparative and analytical uses; for example, the separation of the rare earths is usually achieved by cation exchange followed by elution of their complexes with citric acid.

Electrokinetic phenomena[92]

Electrokinetic is the general description applied to four phenomena which arise when attempts are made to shear off the mobile part of the electric double layer from a charged surface.

If an electric field is applied tangentially along a charged surface, a force is exerted on both parts of the electric double layer. The charged surface (plus attached material) tends to move in the appropriate direction, while the ions in the mobile part of the double layer show a net migration in the opposite direction, carrying solvent along with them, thus causing its flow. Conversely, an electric field is created if the charged surface and the diffuse part of the double layer are made to move relative to each other.

The four electrokinetic phenomena are as follows:

1. *Electrophoresis* – the movement of a charged surface plus attached material (i.e. dissolved or suspended material) relative to stationary liquid by an applied electric field.
2. *Electro-osmosis* – the movement of liquid relative to a stationary charged surface (e.g. a capillary or porous plug) by an applied electric field (i.e. the complement of electrophoresis). The pressure necessary to counterbalance electro-osmotic flow is termed the *electro-osmotic pressure*.
3. *Streaming potential* – the electric field which is created when liquid is made to flow along a stationary charged surface (i.e. the opposite of electro-osmosis).
4. *Sedimentation potential* – the electric field which is created when charged particles move relative to stationary liquid (i.e. the opposite of electrophoresis).

Electrophoresis has the greatest practical applicability of these electrokinetic phenomena and has been studied extensively in its various forms, whereas electro-osmosis and streaming potential have been studied to a moderate extent and sedimentation potential rarely, owing to experimental difficulties.

Electrophoresis[93]

A number of techniques have been developed for studying the migration of colloidal material in an electric field.

Particle (microscope) electrophoresis

If the material under investigation is in the form of a reasonably stable suspension or emulsion containing microscopically visible particles or droplets, then electrophoretic behaviour can be observed and measured directly. Information relevant to soluble material can also be obtained in this way if the substance is adsorbed on to the surface of a carrier, such as oil droplets or silica particles.

The electrophoresis cell usually consists of a horizontal glass tube, of either rectangular or circular cross-section, with an electrode at each end and sometimes with inlet and outlet taps for cleaning and filling (Figures 7.5 and 7.6). Platinum black electrodes are adequate

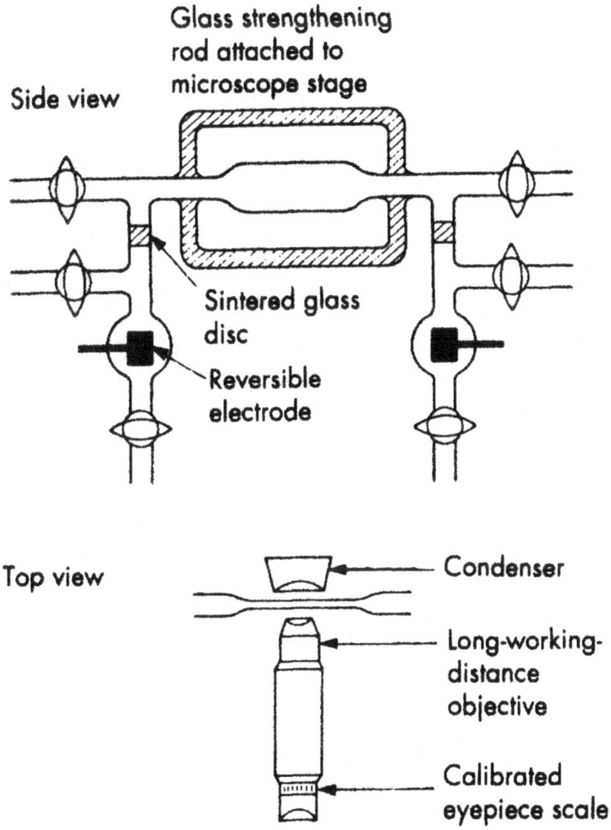

Figure 7.5 A vertically mounted flat particle microelectrophoresis cell[93] (*By courtesy of Academic Press Inc.*)

for salt concentrations below about 0.001 mol dm^{-3} to 0.01 mol dm^{-3}; otherwise appropriate reversible electrodes, such as Cu|$CuSO_4$ or Ag|AgCl, must be used to avoid gas evolution.

Electrophoretic measurements by the microscope method are complicated by the simultaneous occurrence of electro-osmosis. The internal glass surfaces of the cell are usually charged, which causes an electro-osmotic flow of liquid near to the tube walls together with (since the cell is closed) a compensating return flow of liquid with maximum velocity at the centre of the tube. This results in a parabolic distribution of liquid speeds with depth, and the true electrophoretic velocity is only observed at locations in the tube where the electro-osmotic flow and return flow of the liquid cancel. For a cylindrical cell the 'stationary level' is located at 0.146 of the internal diameter from the cell wall. For a flat cell the 'stationary levels' are located at fractions of about 0.2 and 0.8 of the total depth, the exact locations depending on the width/depth ratio. If the particle and cell surfaces have the same zeta potential, the velocity of particles at the centre of the cell is twice their true electrophoretic velocity in a cylindrical cell and 1.5 times their true electrophoretic velocity in a flat cell.

Cylindrical cells are easier to construct and thermostat than flat cells and dark-field illumination can be obtained by the ultramicroscopic method of illuminating the sample perpendicular to the direction of observation (see page 52 and Figure 7.6). The volume of dispersion required is usually less for cylindrical cells than for flat cells and, owing to the relatively small cross-section, it is more often possible to use platinum black rather than reversible electrodes with cylindrical cells. However, unless the capillary wall is extremely thin, an optical correction must be made with cylindrical cells to allow for the focusing action of the tube, and optical distortion may prevent measurements from being made at the far stationary level. Cylindrical cells are unsatisfactory if any sedimentation takes place during the

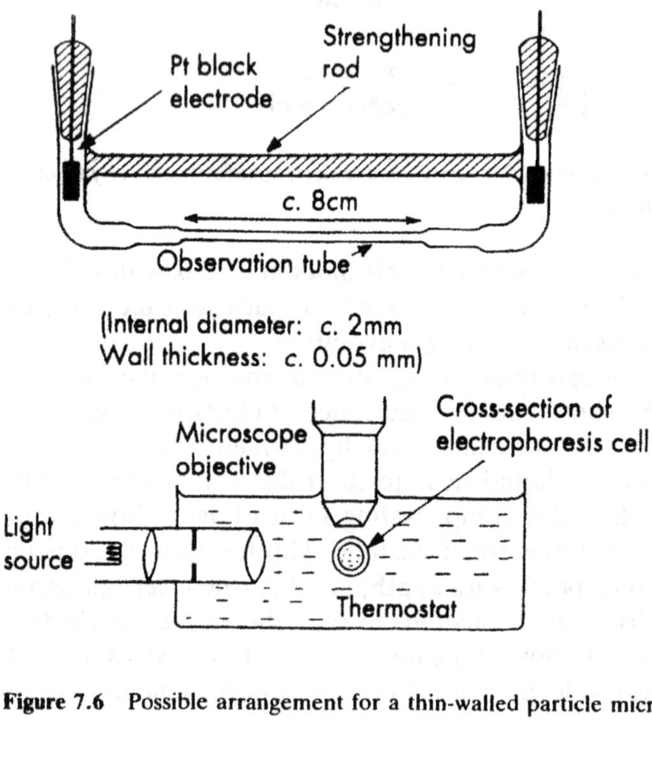

Figure 7.6 Possible arrangement for a thin-walled particle microelectrophoresis cell

measurement; if a rectangular cell is adapted for horizontal viewing (see Figure 7.5), sedimenting particles remain in focus and do not deviate from the stationary levels.

The electrophoretic velocity is found by timing individual particles over a fixed distance (*c.* 100 μm) on a calibrated eyepiece scale. The field strength is adjusted to give timings of *c.* 10 s – faster times introduce timing errors, and slower times increase the unavoidable error due to Brownian motion. Timings are made at both stationary levels. By alternating the direction of the current, errors due to drift (caused by leakage, convection or electrode polarisation) can largely be eliminated. The electrophoretic velocity is usually calculated from the average of the reciprocals of about 20 timings. In a more sophisticated set-up, these timings can be automated using a laser-doppler technique (see page 61).

The potential gradient E at the point of observation is usually calculated from the current I, the cross-sectional area of the channel A and the separately determined conductivity of the dispersion k_0 – i.e. $E = I/k_0A$.

Particle electrophoresis studies have proved to be useful in the investigation of model systems (e.g. silver halide sols and polystyrene latex dispersions) and practical situations (e.g. clay suspensions, water purification, paper-making and detergency) where colloid stability is involved. In estimating the double-layer repulsive forces between particles, it is usually assumed that ψ_d is the operative potential and that ψ_d and ζ (calculated from electrophoretic mobilities) are identical.

Particle electrophoresis is also a useful technique for characterising the surfaces of organisms such as bacteria, viruses and blood cells. The nature of the surface charge can be investigated by studying the dependence of electrophoretic mobility on factors such as pH, ionic strength, addition of specifically adsorbed polyvalent counter-ions, addition of surface-active agents and treatment with specific chemical reagents, particularly enzymes. Figure 7.7 shows, for example, how the mobility–pH curve at constant ionic strength reflects the ionogenic character of some model particle surfaces.

Moving boundary electrophoresis

An alternative electrophoretic technique is to study the movement of a boundary formed between a sol or solution and pure dispersion

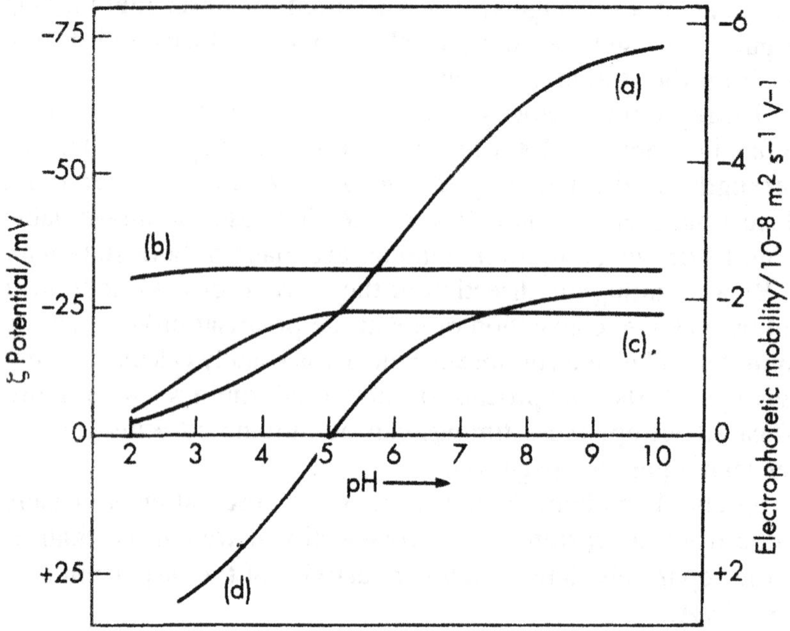

Figure 7.7 Zeta potentials (calculated from electrophoretic mobility data) relating to particles of different ionogenic character plotted as a function of pH in acetate-veronal buffer at constant ionic strength of 0.05 mol dm^{-3}. (a) Hydrocarbon oil droplets. (b) Sulphonated polystyrene latex particles. (c) Arabic acid (carboxylated polymer) adsorbed on to oil droplets. (d) Serum albumin adsorbed on to oil droplets

medium. The Tiselius moving boundary method[184] has found wide application, not only for measuring electrophoretic mobilities, but also for separating, identifying and estimating dissolved macromolecules, particularly proteins. However, as an analytical technique, where electrophoretic mobilities are not required, moving boundary electrophoresis has been largely superseded by simpler and less expensive zone methods.

The Tiselius cell consists of a U-tube of rectangular cross-section, which is divided into a number of sections built up on ground-glass plates so that they can be moved sideways relative to one another. The protein solution is dialysed against buffer (to avoid subsequent disturbance of the boundary by osmotic flow), and then the sections of the cell are filled with buffered protein solution or buffer solution, for example, as shown in Figure 7.8. Large electrode vessels containing reversible electrodes are then attached, and the whole

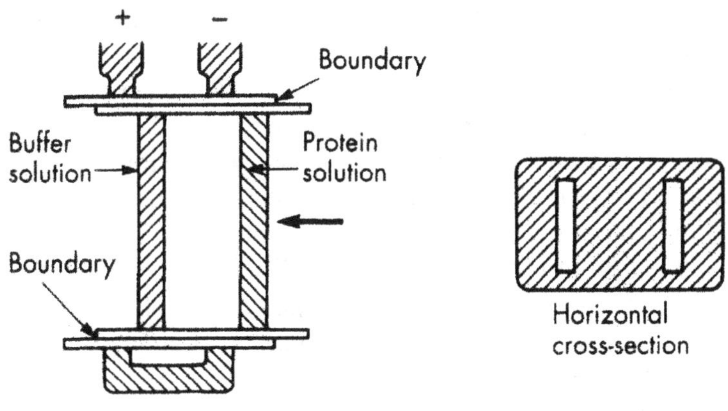

Figure 7.8 A Tiselius electrophoresis cell

assembly is immersed in a thermostat. On attaining hydrostatic and thermal equilibrium, the sections of the cell are slid into alignment to form two sharp boundaries. A current is passed through the cell, and the migration of the boundaries is usually followed by a schlieren technique which shows the boundaries as peaks.

The elongated rectangular cross-section of the cell provides a reasonably long optical path for recording the boundary positions, and at the same time permits efficient thermostatting. By working at around 0–4°C (at which aqueous solutions have maximum density and $d\rho/dT$ is small), convectional disturbance of the boundaries due to the heating effect of the applied current can be minimised even further. The density difference at the boundaries is usually sufficient to prevent disturbance due to electro-osmotic flow at the cell walls.

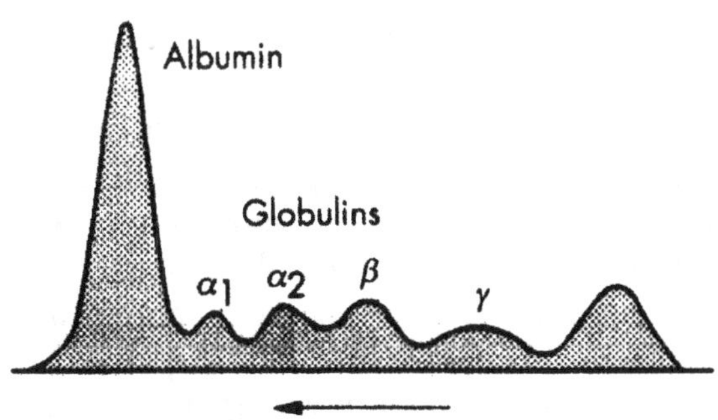

Figure 7.9 Electrophoretic diagram (ascending) for human blood serum

If the protein solution consists of a number of electrophoretically different fractions, the sharp peak corresponding to the initially formed boundary will broaden and may eventually split up into a number of separate peaks each moving at a characteristic speed. The above precautions against boundary disturbances enable a high degree of resolution to be obtained, thus facilitating the identification, characterisation and estimation of the components in such mixtures. For example, the first purpose to which the Tiselius technique was applied[184] was to demonstrate that the component of blood serum once known simply as globulin actually consists of a mixture of several proteins (Figure 7.9).

The moving boundary method is usually complicated by small differences between the ascending and descending boundaries in the two arms of the U-tube. These boundary anomalies result from differences in the conductivity (and, therefore, the potential gradient) at each boundary. They can be minimised by working with low protein concentrations.

Zone electrophoresis[94]

Zone electrophoresis involves the use of a relatively inert and homogeneous solid or gel framework to support the solution under investigation and minimise convectional disturbances. In addition to being experimentally much simpler than moving boundary electrophoresis, it offers the advantages of giving, in principle, complete separation of all electrophoretically different components and of requiring much smaller samples; however, migration through the stabilising medium is generally a complex process and zone electrophoresis is unsuited for the determination of electrophoretic mobilities.

Zone electrophoresis is used mainly as an analytical technique and, to a lesser extent, for small-scale preparative separations. The main applications are in the biochemical and clinical fields, particularly in the study of protein mixtures. Like chromatography, zone electrophoresis is mainly a practical subject, and the most important advances have involved improvements in experimental technique and the introduction and development of a range of suitable supporting media. Much of the earlier work involved the use of filter paper as the supporting medium; however, in recent years filter paper has been somewhat superseded by other materials, such as cellulose acetate, starch gel and polyacrylamide gel, which permit sharper separations.

The particularly high resolving power of moderately concentrated gel media is to a large extent a consequence of molecular sieving acting as an additional separative factor. For example, blood serum can be separated into about 25 components in polyacrylamide gel, but only into 5 components on filter paper or by moving boundary electrophoresis.

Techniques for concentrated dispersions

The electrophoretic mobilities of particles in concentrated dispersion have been measured using (*a*) a relatively simple moving boundary technique[185] and (*b*) a mass transport method[186]. The interpretation of such measurements may be complicated by electric double layer overlap[186].

Streaming current and streaming potential

The development of a streaming potential when an electrolyte is forced through a capillary or porous plug is, in fact, a complex process, charge and mass transfer occurring simultaneously by a number of mechanisms. The liquid in the capillary or plug carries a net charge (that of the mobile part of the electric double layer) and its flow gives rise to a *streaming current* and, consequently, a potential difference. This potential opposes the mechanical transfer of charge by causing backconduction by ion diffusion and, to a much lesser extent, by electro-osmosis. The transfer of charge due to these two effects is called the *leak current*, and the measured *streaming potential* relates to an equilibrium condition when streaming current and leak current cancel each other.

Figure 7.10 illustrates a suitable apparatus for studying streaming potentials. To minimise current drain, a high-resistance voltmeter (most digital voltmeters will suffice), must be used. Most of the difficulties associated with streaming potential measurement originate at the electrodes. A superimposed asymmetry potential often develops; however, by reversing the direction of liquid flow this asymmetry potential can be made to reinforce and oppose the streaming potential and can, therefore, be allowed for.

The streaming current can be measured if the high-resistance voltmeter is replaced with a microammeter of low resistance compared with that of the plug. An alternating streaming current can

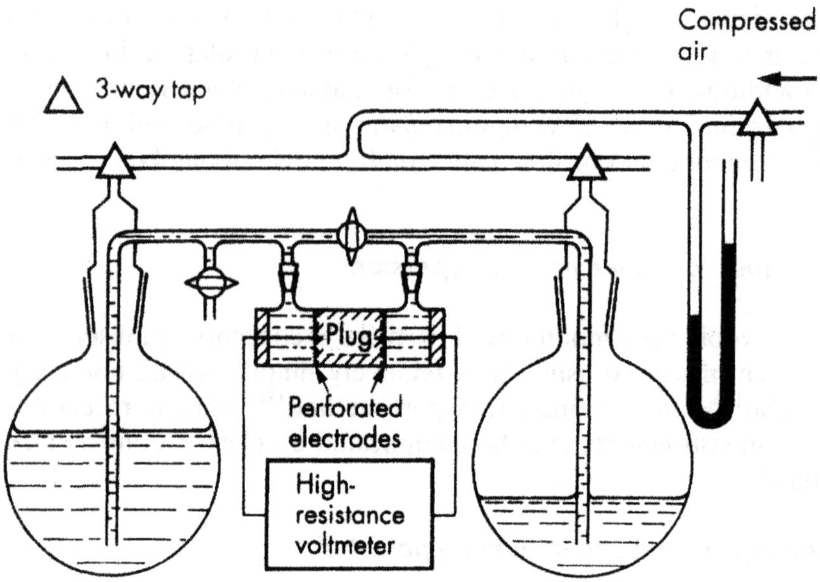

Figure 7.10 A streaming potential apparatus[93] (*By courtesy of Academic Press Inc.*)

be generated by forcing liquid through the plug by means of a reciprocating pump. The main advantage of studying alternating rather than direct streaming currents is that electrode polarisation is far less likely.

Electro-osmosis

Figure 7.11 illustrates a suitable apparatus for studying electro-osmotic flow through a porous plug. Reversible working electrodes are used to avoid gas evolution. A closed system is employed, the electro-osmotic flow rate being determined by measuring the velocity of an air bubble in a capillary tube (*c*. 1 mm diameter) which provides a return path for the electrolyte solution.

It may be necessary to correct the experimental data for effects such as electro-osmosis in the measuring capillary tube and electro-osmotic leak back through the plug.

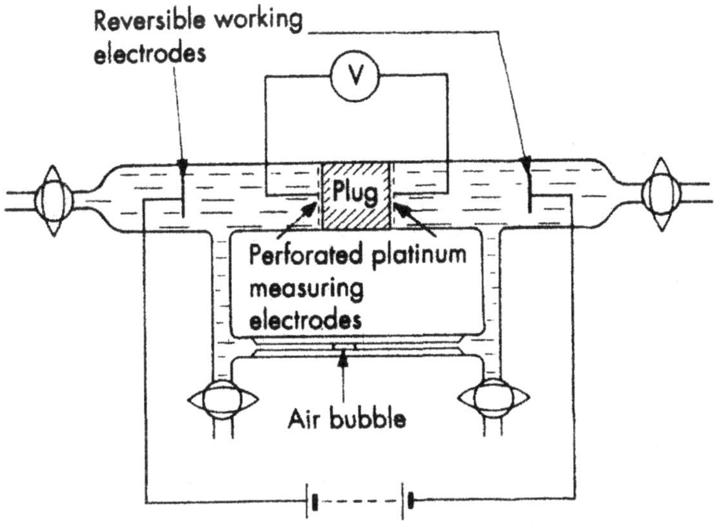

Figure 7.11 An electro-osmosis apparatus[93] (*By courtesy of Academic Press Inc.*)

Electrokinetic theory

Electrokinetic phenomena are only directly related to the nature of the mobile part of the electric double layer and may, therefore, be interpreted only in terms of the zeta potential or the charge density at the surface of shear. No direct information is given about the potentials ψ_0 and ψ_d (although, as already discussed, the value of ζ may not differ substantially from that of ψ_d), or about the charge density at the surface of the material in question.

Electrokinetic theory involves both the theory of the electric double layer and that of liquid flow, and is quite complicated. In this section the relation between electrokinetically determined quantities (particularly electrophoretic mobility) and the zeta potential will be considered.

For curved surfaces the shape of the double layer can be described in terms of the dimensionless quantity 'κa', which is the ratio of radius of curvature to double-layer thickness. When κa is small, a charged particle may be treated as a point charge; when κa is large, the double layer is effectively flat and may be treated as such.

The Hückel equation (small κa)

Consider κa to be small enough for a spherical particle to be treated as a point charge in an unperturbed electric field, but let the particle be large enough for Stokes' law to apply. Equating the electrical force on the particle with the frictional resistance of the medium,

$$Q_E E = 6\pi\eta v_E a$$

or $$u_E = \frac{v_E}{E} = \frac{Q_E}{6\pi\eta a}$$

where Q_E is the net charge on the particle (i.e. the electrokinetic unit), E is the electric field strength, η is the viscosity of the medium, a is the radius of the particle, v_E is the electrophoretic velocity and u_E is the electrophoretic mobility.

The zeta potential is the resultant potential at the surface of shear due to the charges $+Q_E$ of the electrokinetic unit and $-Q_E$ of the mobile part of the double layer – i.e.

$$\zeta = \frac{Q_E}{4\pi\epsilon a} - \frac{Q_E}{4\pi\epsilon\left(a + \frac{1}{\kappa}\right)}$$

$$= \frac{Q_E}{4\pi\epsilon a(1 + \kappa a)} \qquad (7.22)$$

where ϵ is the permittivity of the electrolyte medium (see footnote on page 179). Therefore (neglecting κa compared with unity),

$$u_E = \frac{\zeta\epsilon}{1.5\eta} \qquad (7.23)$$

The Hückel equation is not likely to be applicable to particle electrophoresis in aqueous media; for example, particles of radius 10^{-8} m suspended in a 1–1 aqueous electrolyte solution would require an electrolyte concentration as low as 10^{-5} mol dm^{-3} to give $\kappa a = 0.1$. The equation, however, does have possible applicability to electrophoresis in non-aqueous media of low conductance.

The Smoluchowski equation (large κa)

Consider the motion of liquid in the diffuse part of the double layer relative to that of a non-conducting flat surface when an electric field

E is applied parallel to the surface. Each layer of liquid will rapidly attain a uniform velocity relative and parallel to the surface, with electrical and viscous forces balanced. Equating the electrical and viscous forces on a liquid layer of unit area, thickness dx and distance x from the surface, and having a bulk charge density ρ,

$$E\rho \mathrm{d}x = \left(\eta \frac{\mathrm{d}v}{\mathrm{d}x}\right)_{x+\mathrm{d}x} - \left(\eta \frac{\mathrm{d}v}{\mathrm{d}x}\right)_x$$

$$= \frac{\mathrm{d}}{\mathrm{d}x}\left(\eta \frac{\mathrm{d}v}{\mathrm{d}x}\right)\mathrm{d}x$$

Inserting the Poisson equation in the form $\left[\rho = -\frac{\mathrm{d}}{\mathrm{d}x}\left(\epsilon \frac{\mathrm{d}\psi}{\mathrm{d}x}\right)\right]$

gives

$$-E\frac{\mathrm{d}}{\mathrm{d}x}\left(\epsilon \frac{\mathrm{d}\psi}{\mathrm{d}x}\right) = \frac{\mathrm{d}}{\mathrm{d}x}\left(\eta \frac{\mathrm{d}v}{\mathrm{d}x}\right)$$

Integrating,

$$-E\epsilon \frac{\mathrm{d}\psi}{\mathrm{d}x} = \eta \frac{\mathrm{d}v}{\mathrm{d}x} + \text{constant}$$

The integration constant is zero, since at $x = \infty$, $\mathrm{d}\psi/\mathrm{d}x = 0$ and $\mathrm{d}v/\mathrm{d}x = 0$. Integrating again (assuming that ϵ and η are constant throughout the mobile part of the double layer),

$$-E\epsilon\psi = \eta v + \text{constant}$$

If electrophoresis is being considered, the boundary conditions are $\psi = 0$, $v = 0$ at $x = \infty$ and $\psi = \zeta$, $v = -v_{\mathrm{E}}$ at the surface of shear, where v_{E} is the electrophoretic velocity – i.e. the velocity of the surface relative to stationary liquid. Therefore,

$$E\epsilon\zeta = \eta v_{\mathrm{E}}$$

or

$$u_{\mathrm{E}} = \frac{v_{\mathrm{E}}}{E} = \frac{\zeta\epsilon}{\eta} \tag{7.24}$$

It follows from this expressioin that the electrophoretic mobility of a non-conducting particle for which κa is large at all points on the surface should be independent of its size and shape provided that the zeta potential is constant.

If electro-osmosis is being considered, a similar expression (i.e. $v_{E.O.}/E = \zeta\epsilon/\eta$) is derived, the boundary conditions being $\psi = 0$, $v = v_{E.O.}$ at $x = \infty$ and $\psi = \zeta, v = 0$ at the surface of shear, where $v_{E.O.}$ is the electro-osmotic velocity.

The Henry equation

Henry[187] derived a general electrophoretic equation for conducting and non-conducting spheres which takes the form

$$u_E = \frac{\zeta\epsilon}{1.5\eta}[1 + \lambda F(\kappa a)] \tag{7.25}$$

where $F(\kappa a)$ varies between zero for small values of κa and 1.0 for large values of κa, and $\lambda = (k_0 - k_1)/(2k_0 + k_1)$, where k_0 is the conductivity of the bulk electrolyte solution and k_1 is the conductivity of the particles. For small κa the effect of particle conductance is negligible. For large κa the Henry equation predicts that λ should approach -1 and the electrophoretic mobility approach zero as the particle conductivity increases; however, in most practical cases, 'conducting' particles are rapidly polarised by the applied electric field and behave as non-conductors.

For non-conducting particles ($\lambda = ½$) the Henry equation can be written in the form

$$u_E = \frac{\zeta\epsilon}{1.5\eta} f(\kappa a) \tag{7.26}$$

where $f(\kappa a)$ varies between 1.0 for small κa (Hückel equation) and 1.5 for large κa (Smoluchowski equation) (see Figure 7.12). Zeta potentials calculated from the Hückel equation (for $\kappa a = 0.5$) and from the Smoluchowski equation (for $\kappa a = 300$) differ by about 1 per cent from the corresponding zeta potentials calculated from the Henry equation.

The Henry equation is based on several simplifying assumptions:

1. The Debye–Hückel approximation is made.
2. The applied electric field and the field of the electric double layer

are assumed to be simply superimposed. Mutual distortion of these fields could affect electrophoretic mobility in two ways: (*a*) through abnormal conductance (*surface conductance*) in the vicinity of the charged surface, and (*b*) through loss of double-layer symmetry (*relaxation effect*).

3. ϵ and η are assumed to be constant throughout the mobile part of the double layer.

Surface conductance

The distribution of ions in the diffuse part of the double layer gives rise to a conductivity in this region which is in excess of that in the bulk electrolyte medium. Surface conductance will affect the distribution of electric field near to the surface of a charged particle and so influence its electrokinetic behaviour. The effect of surface conductance on electrophoretic behaviour can be neglected when κa is small, since the applied electric field is hardly affected by the particle in any case. When κa is not small, calculated zeta potentials may be significantly low, on account of surface conductance.

According to Booth and Henry[188], the equation relating electrophoretic mobility with zeta potential for non-conducting spheres with large κa when corrected for surface conductance takes the form

$$u_{\mathrm{E}} = \frac{\zeta\epsilon}{\eta}\left[\frac{k_0}{k_0 + (k_{\mathrm{s}}/a)}\right] \tag{7.27}$$

where k_0 is the conductivity of the bulk electrolyte medium and k_{s} is the surface conductivity.

Substituting ζ_{a} for $\eta u_{\mathrm{E}}/\epsilon$ (i.e ζ_{a} is the apparent zeta potential calculated from the Smoluchowski equation) gives

$$\frac{1}{\zeta_{\mathrm{a}}} = \frac{1}{\zeta}\left(1 + \frac{k_{\mathrm{s}}}{k_0 a}\right) \tag{7.28}$$

A plot of $1/\zeta_{\mathrm{a}}$ against $1/a$ should, therefore, give a straight line (if κa is large and if k_{s}, k_0 and ζ are constant) from which a zeta potential corrected for surface conductance can be obtained by extrapolation. Zeta potentials for oil droplets and protein-covered glass particles have been determined in this way[189].

The importance of surface conductance at large κa clearly depends on the magnitude of $k_{\mathrm{s}}/(k_0 a)$ compared with unity. The surface conductivity in the mobile part of the double layer can be calculated (and is allowed for in the treatments of relaxation which are outlined in the next section). Experimental surface conductivities (which are not very reliable) tend to be higher than those calculated for the mobile part of the double layer, and the possibility of surface conductance inside the shear plane, especially if the particle surface is porous, has been suggested to account for this discrepancy[90,188]. There is, therefore, some uncertainty regarding the influence of surface conductance on electrophoretic behaviour; however, it is unlikely to be important when the electrolyte concentration is greater than *c.* 0.01 mol dm^{-3}.

Relaxation

The ions in the mobile part of the double layer show a net movement in a direction opposite to that of the particle under the influence of the applied electric field. This creates a local movement of liquid which opposes the motion of the particle, and is known as *electrophoretic retardation*. It is allowed for in the Henry equation.

The movement of the particle relative to the mobile part of the double layer results in the double layer being distorted, because a finite time (relaxation time) is required for the original symmetry to be restored by diffusion and conduction. The resulting asymmetric mobile part of the double layer exerts an additional retarding force on the particle, known as the *relaxation effect*, and this is not accounted for in the Henry equation. Relaxation can be safely neglected when κa is either small (< *c.* 0.1) or large (> *c.* 300), but it is significant for intermediate values of κa especially at high potentials and when the counter-ions are of high charge number and/or have low mobilities.

Wiersema, Loeb and Overbeek[190] have derived equations which allow for retardation, relaxation and for surface conductance in the mobile part of the double layer, and have solved them numerically by computer. The main assumptions upon which this treatment is based are:

1. The particle is a rigid, non-conducting sphere with its charge uniformly distributed over the surface.
2. The electrophoretic behaviour of the particle is not influenced by other particles in the dispersion.

3. Permittivity and viscosity are constant throughout the mobile part of the double layer, which is described by the classical Gouy–Chapman theory.
4. Only one type each of positive and negative ions are present in the mobile part of the double layer.

Figures 7.12 and 7.13 show the result of a selection of these computations.

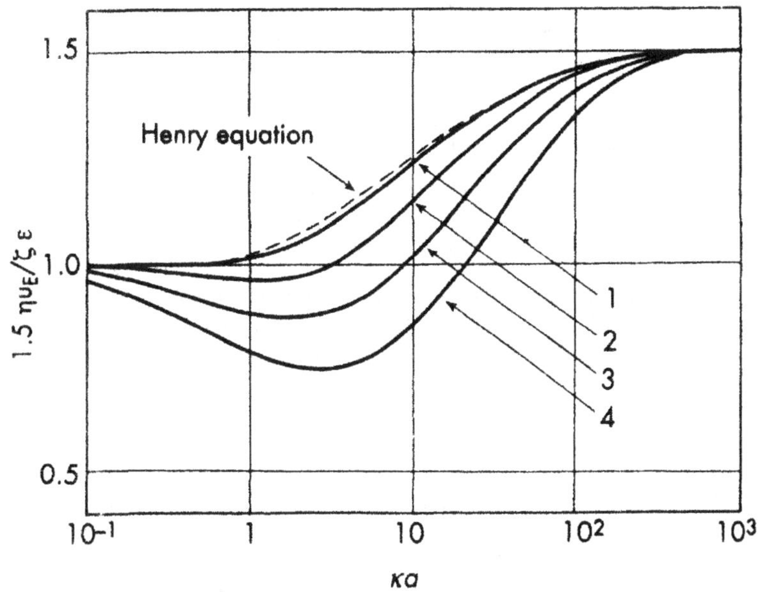

Figure 7.12 Electrophoretic mobility and zeta potential for spherical colloidal particles in 1–1 electrolyte solutions ($\Lambda_+ = \Lambda_- = 70\ \Omega^{-1}\ \mathrm{cm}^2\ \mathrm{mol}^{-1}$). The curves refer to $e\zeta/kT$ = 1, 2, 3 and 4 (i.e. ζ/mV = 25.6, 51.2, 76.8 and 102.4 at 25°C) [After P.H. Wiersema, A.L. Loeb and J.Th.G. Overbeek, *J. Colloid Interface Sci.*, **22**, 78 (1966)]

Investigations of the electrophoretic behaviour of monodispersed carboxylated polystyrene latex dispersions as a function of particle size and electrolyte concentration by Shaw and Ottewill[191] have confirmed, at least qualitatively, the existence of κa and relaxation effects.

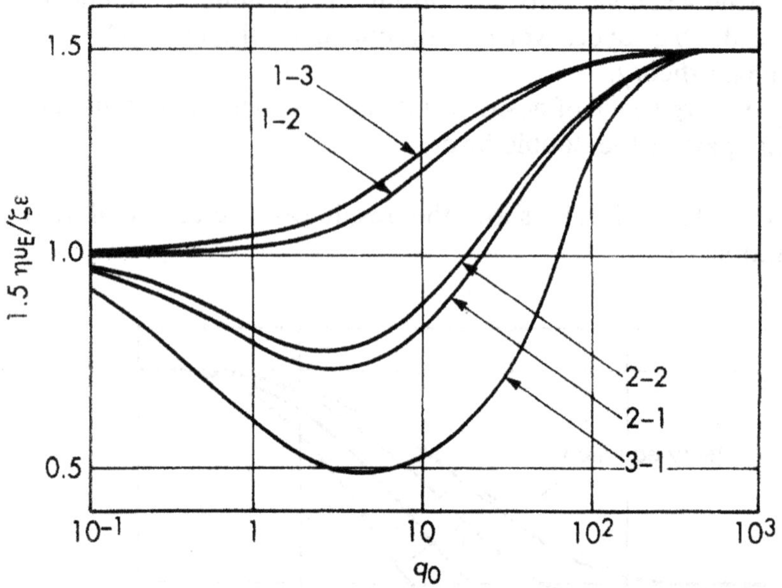

Figure 7.13 Electrophoretic mobility and zeta potential for spherical colloidal particles in electrolyte solutions containing polyvalent ions ($\Lambda_+/z_+ = \Lambda_-/z_- = 70\ \Omega^{-1}$ $cm^2\ mol^{-1}$). Electrolyte type is numbered with counter-ion charge number first:

$$e\zeta/kT = 2 \text{ in each case}$$
$$q_0 = (2e^2N_Acz^2/\varepsilon kT)^{1/2}a$$

where c is the electrolyte concentration and z is the counter-ion charge number [After P.H. Wiersema, A.L. Loeb and J.Th.G. Overbeek, *J. Colloid Interface Sci.*, **22**, 78 (1966)]

Permittivity and viscosity

Further difficulty in the calculation and interpretation of zeta potentials will arise if the electric field strength ($d\psi/dx$) close to the shear plane is high enough to significantly decrease ϵ and/or increase η by dipole orientation. Lyklema and Overbeek[192] examined this problem and concluded that the effect of $d\psi/dx$ on ϵ is insignificant, but that its effect on η may be significant, especially at high potential and high electrolyte concentration. A significant (and positive) viscoelectric effect would result in the effective location of the shear plane moving farther away from the particle surface with increasing ζ and/or increasing κ; in other words, the physical meaning of the term 'zeta potential' would vary. More recently investigations by Stigter and Hunter[193] suggest, however, that the viscoelectric effect was overestimated by Lyklema and Overbeek and that it is, in fact, insignificant in most practical situations.

Streaming current and streaming potential

The classical equations relating streaming current or streaming potential to zeta potential are derived for the case of a single circular capillary as follows.

Let E_S be the potential difference developed between the ends of a capillary tube of radius a and length l for an applied pressure difference p. Assuming laminar flow, the liquid velocity v_x at a distance x measured from the surface of shear and along a radius of the capillary is given by Poiseuille's equation, which can be written in the form

$$v_x = \frac{p(2ax - x^2)}{4\eta l}$$

The volume of liquid with velocity v_x can be represented by a hollow cylinder of radius $(a-x)$ and thickness dx. The rate of flow, dJ, in this cylindrical layer is, therefore, given by

$$dJ = 2\pi(a - x)v_x dx = \frac{2\pi p(2ax - x^2)(a - x)dx}{4\eta l}$$

The streaming current I_s is given by

$$I_s = \int_0^a \rho dJ$$

where ρ is the bulk charge density. If κa is large, the potential decay in the double layer and, therefore, the streaming current are located in a region close to the wall of the capillary tube where x is small compared with a. Substituting for ρ (Poisson's equation, $d^2\psi/dx^2\ -\rho/\epsilon$) and dJ (neglecting x compared with a) gives

$$I_s = \frac{\pi\epsilon p a^2}{\eta l}\int_0^a x\frac{d^2\psi}{dx^2}dx$$

The solution of this expression (by partial integration), with the boundary conditions ($\psi = \zeta$ at $x = 0$ and $\psi = 0$, $d\psi/dx = 0$ at $x = a$) taken into account, is

$$I_s = \frac{\epsilon p A \zeta}{\eta l} \tag{7.29}$$

where A is the cross-sectional area of the capillary.

The streaming current and the streaming potential are related by Ohm's law,

$$E_s = \frac{I_s l}{k_0 A}$$

where k_0 is the conductivity of the electrolyte solution. Therefore,

$$E_s = \frac{\epsilon p \zeta}{\eta k_0} \tag{7.30}$$

k_0 can be corrected to include a surface conductivity term k_s and equation (7.30) becomes

$$E_s = \frac{\epsilon p \zeta}{\eta\left(k_0 + \frac{2k_s}{a}\right)} \tag{7.31}$$

A more general derivation[90] for porous plugs also leads to equations (7.29) and (7.30). However, for porous plugs, there is no satisfactory method of correcting streaming potential data for surface conductance. If equation (7.31) is used with a equal to the average pore radius, the calculated zeta potentials are too low. The importance of surface conductance can be investigated qualitatively by comparing the conductivity ratio of two relevant electrolyte concentrations in bulk and in the plug. A knowledge of the surface conductance is not required when relating streaming current to zeta potential. The situation in porous plugs may also be more complicated than accounted for above if (*a*) the effective area of the plug for the streaming current differs from that for the leak current as a result of the different mechanisms involved, and (*b*) the plug is compressible and the applied pressure affects the average pore size. The validity of all zeta potentials calculated from streaming (and electro-osmotic) measurements on porous plugs is somewhat dubious.

Electro-osmosis

Experimentally, it is the volume flow rate which is measured. For a single capillary of cross-sectional area A and with large κa, this flow rate is given from the Smoluchowski equation by

$$\frac{dV_{E.O.}}{dt} = Av_{E.O.} = \frac{AE\epsilon\zeta}{\eta}$$

and since (by Ohm's law) $AE = I/k_0$, where k_0 is the conductivity of the liquid and I the current,

$$\frac{dV_{E.O.}}{dt} = \frac{\epsilon I\zeta}{\eta k_0} \tag{7.32}$$

or, correcting k_0 to account for the surface conductivity k_s,

$$\frac{dV_{E.O.}}{dt} = \frac{\epsilon I\zeta}{\eta\left(k_0 + \frac{2k_s}{a}\right)} \tag{7.33}$$

Equation (7.32) can be shown by a more general derivation[90] to apply also to porous plugs, but, as in the case of streaming potential, there is no satisfactory method of allowing for surface conductance.

Combination of equations (7.30) and (7.32) gives a general relationship between electro-osmosis and streaming potential, Saxen's equation, which, by a more general derivation, can be shown to apply independent of plug structure and surface conductance:

$$\frac{dV_{E.O.}/dt}{I} = \frac{E_s}{p} \tag{7.34}$$

8 *Colloid stability*

A most important physical property of colloidal dispersions is the tendency of the particles to aggregate. Encounters between particles dispersed in liquid media occur frequently and the stability of a dispersion is determined by the interaction between the particles during these encounters.

The principal cause of aggregation is the van der Waals attractive forces between the particles, which are long-range forces. To counteract these and promote stability, equally long-range repulsive forces are required. Solvation tends to be too short-range; however, the molecular ordering associated with solvation can propagate several molecular diameters into the liquid phase and may exert some influence on stability[95]. The principal stabilising options are electrostatic (i.e. the overlap of similarly charged electric double layers) and polymeric. Polymeric and/or surfactant additives can influence stability by a variety of mechanisms and the overall situation is often very complicated.

Lyophobic sols

Ideally, lyophobic sols are stabilised entirely by electric double-layer interactions and, as such, present colloid stability at its simplest.

Critical coagulation concentrations* – Schulze–Hardy rule

A most notable property of lyophobic sols is their sensitivity to coagulation by small amounts of added electrolyte. The added

*In this book, *coagulation* refers to a primary minimum effect (see page 219) and *flocculation* refers to a secondary minimum or polymer bridging effect (see pages 220, 238 and 241). In much of the colloid science literature, these terms (together with the generic term *aggregation*) are used interchangeably.

Table 8.1 Critical coagulation concentrations (in millimoles per dm^3) for hydrophobic sols[96] *(By courtesy of Elsevier Publishing Company)*

As_2S_3 (−ve sol)		AgI (−ve sol)		Al_2O_3 (+ve sol)	
$LiCl$	58	$LiNO_3$	165	$NaCl$	43.5
$NaCl$	51	$NaNO_3$	140	KCl	46
KCl	49.5	KNO_3	136	KNO_3	60
KNO_3	50	$RbNO_3$	126		
K acetate	110	$AgNO_3$	0.01)		
$CaCl_2$	0.65	$Ca(NO_3)_2$	2.40	K_2SO_4	0.30
$MgCl_2$	0.72	$Mg(NO_3)_2$	2.60	$K_2Cr_2O_7$	0.63
$MgSO_4$	0.81	$Pb(NO_3)_2$	2.43	K_2 oxalate	0.69
$AlCl_3$	0.093	$Al(NO_3)_3$	0.067	$K_3[Fe(CN)_6]$	0.08
$\frac{1}{2}Al_2(SO_4)_3$	0.096	$La(NO_3)_3$	0.069		
$Al(NO_3)_3$	0.095	$Ce(NO_3)_3$	0.69		

electrolyte causes a compression of the diffuse parts of the double layers around the particles and may, in addition, exert a specific effect through ion adsorption into the Stern layer. The sol coagulates when the range of double-layer repulsive interaction is sufficiently reduced to permit particles to approach close enough for van der Waals forces to predominate.

The critical coagulation concentration (c.c.c.) of an indifferent (inert) electrolyte (i.e. the concentration of the electrolyte which is just sufficient to coagulate a lyophobic sol to an arbitrarily defined extent in an arbitrarily chosen time) shows considerable dependence upon the charge number of its counter-ions. In contrast, it is practically independent of the specific character of the various ions, the charge number of the co-ions and the concentration of the sol, and only moderately dependent on the nature of the sol. These generalisations are illustrated in Table 8.1, and are known as the Schulze–Hardy rule.

The Deryagin–Landau and Verwey–Overbeek theory (introduction)

Deryagin and Landau[194] and Verwey and Overbeek[97] independently developed a quantitative theory in which the stability of lyophobic sols, especially in relation to added electrolyte, is treated in terms of the energy changes which take place when particles approach one another. The theory involves estimations of the energy due to the overlap of electric double layers (usually repulsion) and the London–van der Waals energy (usually attraction) in terms of interparticle distance, and their summation to give the total interaction energy in terms of interparticle distance. Colloid stability is then interpreted in terms of the nature of the interaction energy–distance curve (see Figures 8.2–8.4). Theoretical calculations have been made for the interactions (*a*) between two parallel charged plates of infinite area and thickness, and (*b*) between two charged spheres. The calculations for the interaction between flat plates are relevant to the stability of thin soap films, and have been related with a reasonable measure of success to experimental studies in this field[98] (see Chapter 10). The calculations for the interaction between spheres are relevant to the stability of dispersions and will be outlined. In fact, the conclusions arising from both theoretical treatments are broadly similar.

Double-layer interaction energies

The calculation of the interaction energy, V_R, which results from the overlapping of the diffuse parts of the electric double layers around two spherical particles (as described by Gouy–Chapman theory) is complex. No exact analytical expression can be given and recourse must be had to numerical solutions or to various approximations.

If it is assumed that ion adsorption equilibrium is maintained as two charged particles approach each other and their double layers overlap, two well-defined situations can be recognised. If the surface charge is the result of the adsorption of potential-determining ions, the surface potential remains constant and the surface charge density adjusts accordingly; but if the surface charge is the result of ionisation, the surface charge density remains constant and the surface potential adjusts accordingly (see page 180). At large interparticle separations the difference between constant potential and constant charge interactions will be minimal. Overbeek[99] has considered this problem and concluded that the rate of double-layer overlap in a typical Brownian motion encounter between particles is too fast for adsorption equilibrium to be maintained and that the true situation will, in general, lie somewhere between constant potential and constant charge.

For the case of two spherical particles of radii a_1 and a_2, Stern potentials, ψ_{d1} and ψ_{d2}, and a shortest distance, H, between their Stern layers, Healy and co-workers[195] have derived the following expressions for constant-potential, V_R^ψ, and constant-charge, V_R^σ, double-layer interactions. The low-potential form of the Poisson–Boltzmann distribution (equation 7.12) is assumed to hold and κa_1 and κa_2 are assumed to be large compared with unity:

$$V_R^\psi = \frac{\pi\epsilon a_1 a_2 (\psi_{d1}^2 + \psi_{d2}^2)}{(a_1 + a_2)} \left\{ \frac{2\psi_{d1}\psi_{d2}}{(\psi_{d1}^2 + \psi_{d2}^2)} \cdot \ln\left(\frac{1+\exp[-\kappa H]}{1-\exp[-\kappa H]}\right) + \ln(1-\exp[-2\kappa H]) \right\} \tag{8.1}$$

$$V_R^\sigma = \frac{\pi\epsilon a_1 a_2 (\psi_{d1}^2 + \psi_{d2}^2)}{(a_1 + a_2)} \left\{ \frac{2\psi_{d1}\psi_{d2}}{(\psi_{d1}^2 + \psi_{d2}^2)} \cdot \ln\left(\frac{1+\exp[-\kappa H]}{1-\exp[-\kappa H]}\right) - \ln(1-\exp[-2\kappa H]) \right\} \tag{8.2}$$

where ϵ is the permittivity of the dispersion medium and κ is as defined in equation (7.6)

Table 8.2 shows the signs of V_R that accord with equations (8.1) and (8.2) for different homocoagulation and heterocoagulation situations. (N.B. Attraction is negative and repulsion positive.)

For equal spheres, with $a_1 = a_2 = a$ and $\psi_{d1} = \psi_{d2} = \psi_d$, equations (8.1) and (8.2) reduce to

$$V_R^\psi = 2\pi\epsilon a\psi_d^2 \ln(1+\exp[-\kappa H]) \tag{8.3}$$

and

$$V_R^\sigma = 2\pi\epsilon a\psi_d^2 \ln(1+\exp[-\kappa H]) \tag{8.4}$$

For small electric double layer overlap, such that $\exp[-\kappa H] \ll 1$, these expressions both reduce to

$$V_R = 2\pi\epsilon a\psi_d^2 \exp[-\kappa H] \tag{8.5}$$

Table 8.2 Predicted signs of V_R*

Situation	V_R^{ψ}	V_R^{σ}
(a) $\psi_{d1} = \psi_{d2} \neq 0$	+ve	+ve
(b) ψ_{d1} and ψ_{d2} of like sign but unequal magnitude	+ve at large H −ve at small H	+ve
(c) ψ_{d1} and ψ_{d2} of opposite sign	−ve	−ve at large H +ve at small H
(d) ψ_{d1} or $\psi_{d2} = 0$	−ve	+ve

Another approximate expression for V_R is that given by Reerink and Overbeek[196]. The Debye–Hückel low-potential approximation is not made, but the interparticle distance is considered to be sufficiently large (i.e. exp $[-\kappa H] \ll 1$) for the potential at any point between the particles to be given by the sum of the individual potentials at that point for each particle in the absence of the other. For unequal spherical particles,

$$V_R = \frac{64\pi\varepsilon a_1 a_2 k^2 T^2 \gamma_1 \gamma_2}{(a_1 + a_2)e^2 z^2} \exp[-\kappa H] \tag{8.6}$$

which, for equal spheres, reduces to

$$V_R = \frac{32\pi\varepsilon a k^2 T^2 \gamma^2}{e^2 z^2} \exp[-\kappa H] \tag{8.7}$$

where z is the counter-ion charge number and

$$\gamma = \frac{\exp[ze\psi_d / 2kT] - 1}{\exp[ze\psi_d / 2kT] + 1} \tag{7.5}$$

If the Debye–Hückel approximation, $ze\psi_d/kT \ll 1$, is made, equation (8.7) reduces to equation (8.5).

So far, only non-specific ion adsorption in the diffuse part of the electric double layer has been considered. The broad prediction is that V_R should decrease in an approximately exponential fashion with increasing H and that the range of V_R should be decreased by

*The symbols V_R and V_A are those traditionally used to represent electric double layer and van der Waals interactions, respectively. The subscripts, R and A, reflect the usual, but not universal, repulsive and attractive nature of these interactions.

increasing κ (i.e. by increasing electrolyte concentration and/or counter-ion charge number). Specific effects may also influence V_R[100]. Counter-ion adsorption in the Stern layer may cause a reversal of charge (see page 183), so that V_R for a pair of identical particles will be zero at the reversal of charge concentration and positive (repulsion) at both below and above this concentration. In contrast to the effect of electrolyte on the diffuse part of the electric double layer, the amount of added electrolyte required to produce such a specific effect will depend on the total surface area of the particles. The nature of the electric double layer (and of V_R) may also be influenced by ion hydrolysis and/or complexation reactions[100–101].

An interesting example of electrostatic attraction of oppositely charged surfaces is that exhibited by kaolinite clay particles[18]. The faces of the plate-like particles tend to be negatively charged and the edges positively charged. This can be demonstrated by introducing negatively charged colloidal gold particles into the clay suspension, then subsequently taking an electron micrograph, which shows the small gold particles adhering to the edges (but not to the faces) of the clay platelets. Edge-to-face attraction between the clay platelets can lead to the formation of a 'cardhouse' structure with a relatively low particle density.

van der Waals forces between colloidal particles

The forces of attraction between neutral, chemically saturated molecules, postulated by van der Waals to explain non-ideal gas behaviour, also originate from electrical interactions. Three types of such intermolecular attraction are recognised:

1. Two molecules with permanent dipoles mutually orientate each other in such a way that, on average, attraction results.
2. Dipolar molecules induce dipoles in other molecules so that attraction results.
3. Attractive forces are also operative between non-polar molecules, as is evident from the liquefaction of hydrogen, helium, etc. These universal attractive forces (known as *dispersion forces*) were first explained by London (1930) and are due to the polarisation of one molecule by fluctuations in the charge distribution in a second molecule, and vice versa.

With the exception of highly polar materials, London dispersion forces account for nearly all of the van der Waals attraction which is operative. The London attractive energy between two molecules is very short-range, varying inversely with the sixth power of the intermolecular distance. For an assembly of molecules, dispersion forces are, to a first approximation, additive and the van der Waals interaction energy between two particles can be computed by summing the attractions between all interparticle molecule pairs.

The results of such summations predict that the London interaction energy between collections of molecules (e.g. between colloidal particles) decays much less rapidly than that between individual molecules.

For the case of two spherical particles of radii a_1 and a_2, separated *in vacuo* by a shortest distance H, Hamaker[197] derived the following expression for the London dispersion interaction energy, V_A:

$$V_A = -\frac{A}{12}\left[\frac{y}{x^2 + xy + x} + \frac{y}{x^2 + xy + x + y} + 2\ln\left(\frac{x^2 + xy + x}{x^2 + xy + x + y}\right)\right] \tag{8.8}$$

where

$$x = \frac{H}{a_1 + a_2} \text{ and } y = a_1 / a_2$$

A is a constant, known as the Hamaker constant.

For equal spheres, with $a_1 = a_2 = a$ (i.e. $x = H/2a$), equation (8.8) takes the form

$$V_A = -\frac{A}{12}\left[\frac{1}{x(x+2)} + \frac{1}{(x+1)^2} + 2\ln\left(\frac{x(x+2)}{(x+1)^2}\right)\right] \tag{8.9}$$

If a small interparticle separation is assumed, such that $H \ll a$ (i.e. $x \ll 1$), this rather awkward equation simplifies to

$$V_A = -\frac{A}{12}\cdot\frac{1}{2x} = -\frac{Aa}{12H} \tag{8.10}$$

Values of V_A calculated from this equation will be overestimated on account of the above approximation.

Values of V_A calculated from any of the above equations will be overestimated at large distances ($H >$ c. 10 nm) owing to a neglect of the finite time required for propagation of electromagnetic radiation between the particles, the result of which is a weakening of V_A. In most practical situations relating to colloid stability this *retardation effect* is not likely to be important.

The major problem in calculating the van der Waals interaction between colloidal particles is that of evaluating the Hamaker constant, A. Two methods are available.

The first of these methods is the London–Hamaker microscopic approach, which has already been mentioned. In it Hamaker constants are evaluated from the individual atomic polarisabilities and the atomic densities of the materials involved. The total interaction is assumed to be the sum of the interactions between all interparticle atom pairs and is assumed to centre around a single oscillation frequency. These assumptions are essentially incorrect. The influence of neighbouring atoms on the interaction of a given pair of atoms is ignored. van der Waals interaction energies calculated in accord with the microscopic approach are likely to be in error but the error involved is not likely to be so great as to prejudice general conclusions concerning colloid stability.

The other method is the macroscopic approach of Lifshiftz[95,102,103,198,199], in which the interacting particles and the intervening medium are treated as continuous phases. The calculations are complex, and require the availability of bulk optical/dielectric properties of the interacting materials over a sufficiently wide frequency range.

The values of A calculated by microscopic and by macroscopic methods tend to be similar in the non-retarded range. The macroscopic approach predicts a smaller retardation effect (i.e. better applicability of equations 8.8–8.10 for relatively large values of H) than the microscopic approach[104].

Hamaker constants for single materials usually vary between about 10^{-20} J and 10^{-19} J. Some examples are given in Table 8.3. Where a range of values is quoted for a given material, this reflects different methods of calculation within the basic microscopic or macroscopic method.

The presence of a liquid dispersion medium, rather than a vacuum (or air), between the particles (as considered so far) notably lowers the van der Waals interaction energy. The constant A in equations (8.8)–(8.10) must be replaced by an effective Hamaker constant. Consider the interaction between two particles, 1 and 2, in a dispersion medium, 3. When the particles are far apart (Figure 8.1a),

Table 8.3 Values of Hamaker constants[105]

Material	A_{11} (*microscopic*)	A_{11} (*macroscopic*)
	10^{-20} J	10^{-20} J
Water	3.3– 6.4	3.0– 6.1
Ionic crystals	15.8–41.8	5.8–11.8
Metals	7.6–15.9	22.1
Silica	50	8.6
Quartz	11.0–18.6	8.0– 8.8
Hydrocarbons	4.6–10	6.3
Polystyrene	6.2–16.8	5.6– 6.4

Figure 8.1

the interactions are particle–dispersion medium interactions, with Hamaker constants A_{13} and A_{23}. If particle 2 is brought close to particle 1 (Figure 8.1b), dispersion medium must be displaced to the position originally occupied by particle 2 and the above interactions are replaced by particle–particle and dispersion medium–dispersion medium interactions, with Hamaker constants A_{12} and A_{33}. The effective Hamaker constant is, therefore, given by

$$A_{132} = A_{12} + A_{33} - A_{13} - A_{23} \tag{8.11}$$

If the attractions between unlike phases is taken to be the geometric mean of the attractions of each phase to itself (cf. equation 4.3) – i.e. $A_{12} = (A_{11} \times A_{22})^{1/2}$, $A_{13} = (A_{11} \times A_{33})^{1/2}$ and $A_{23} = (A_{22} \times A_{33})^{1/2}$ then equation (8.11) becomes

$$A_{132} = (A_{11}^{1/2} - A_{33}^{1/2})(A_{22}^{1/2} - A_{33}^{1/2}) \tag{8.12}$$

If the two particles are of the same material, this expression becomes

$$A_{131} = (A_{11}^{1/2} - A_{33}^{1/2})^2 \tag{8.13}$$

giving values of A_{131} for hydrosols of up to about 10^{-19} J.

A_{132} will be positive (interparticle attraction) where A_{11} and A_{22} are either both greater than or both less than A_{33}. However, in the unusual situation where A_{33} has a value intermediate between those of A_{11} and A_{22}, then A_{132} is negative – i.e. a repulsive van der Waals interaction between the particles is predicted.

A_{131} for the interaction of particles of the same material is always positive – i.e. the van der Waals interaction energy is always one of attraction. This interaction will be weakest when the particles and the dispersion medium are chemically similar, since A_{11} and A_{33} will be of similar magnitude and the value of A_{131} will therefore, be low.

Potential energy curves

The total energy of interaction between the particles in a lyophobic sol is obtained by summation of the electric double layer and van der Waals energies, as illustrated in Figure 8.2.

The general character of the resulting potential energy–distance curve can be deduced from the properties of the two components. For the interaction between particles of the same material, the double-layer repulsion energy (equation (8.5)) is an approximately exponential function of the distance between the particles with a range of the order of the thickness of the double layer ($1/\kappa$), and the van der Waals attraction energy (equation (8.10)) decreases as an inverse power of the distance between the particles. Consequently, van der Waals attraction will predominate at small* and at large interparticle distances. At intermediate distances double-layer repulsion may predominate, depending on the actual values of the two forces. Figure 8.2 shows the two general types of potential energy curve which are possible. The total potential energy curve $V(1)$ shows a repulsive energy maximum, whereas in curve $V(2)$ the double-layer repulsion does not predominate over van der Waals attraction at any interparticle distance.

*Repulsion due to overlapping of electron clouds (Born repulsion) predominates at very small distances when the particles come into contact, and so there is a deep minimum in the potential energy curve which is not shown in Figures 8.2–8.4.

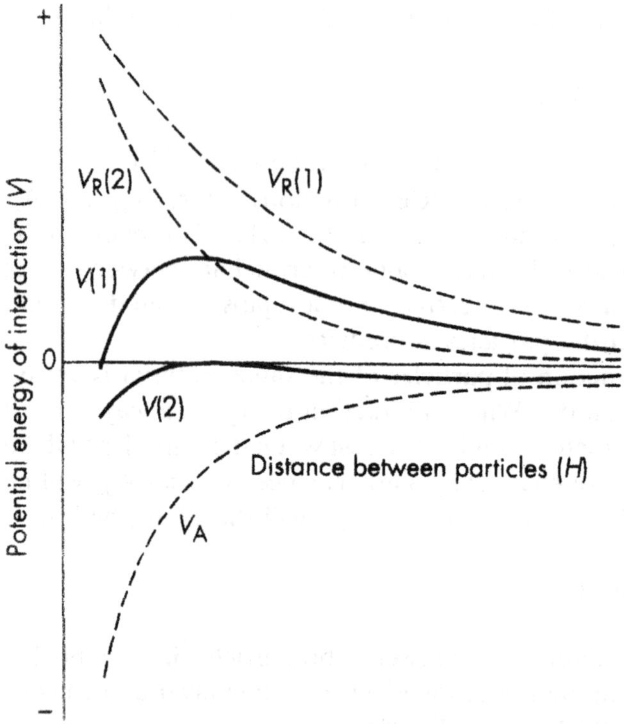

Figure 8.2 Total interaction energy curves, $V(1)$ and $V(2)$, obtained by the summation of an attraction curve, V_A, with different repulsion curves, $V_R(1)$ and $V_R(2)$

If the potential energy maximum is large compared with the thermal energy kT of the particles, the system should be stable; otherwise, the system should coagulate. The height of this energy barrier to coagulation depends upon the magnitude of ψ_d (and ζ) and upon the range of the repulsive forces (i.e. upon $1/\kappa$), as shown in Figures 8.3 and 8.4.

Figures 8.3 and 8.4 show potential energy maxima at inter-particle separations of a few nanometres. Surface roughness of at least up to this magnitude is, therefore, unlikely to invalidate these potential energy calculations.

Secondary minima

Another characteristic feature of these potential energy curves is the presence of a secondary minimum at relatively large interparticle distances. If this minimum is moderately deep compared with kT, it should give rise to a loose, easily reversible flocculation. For small particles ($a < c.\ 10^{-8}$ m) the secondary minimum is never deep enough for this to happen in those cases where the potential energy maximum is high enough to prevent coagulation into the primary minimum. If the particles are larger, flocculation in the secondary minimum may cause observable effects.

Several colloidal systems containing anisodimensional particles, such as iron oxide and tobacco mosaic virus sols, show reversible

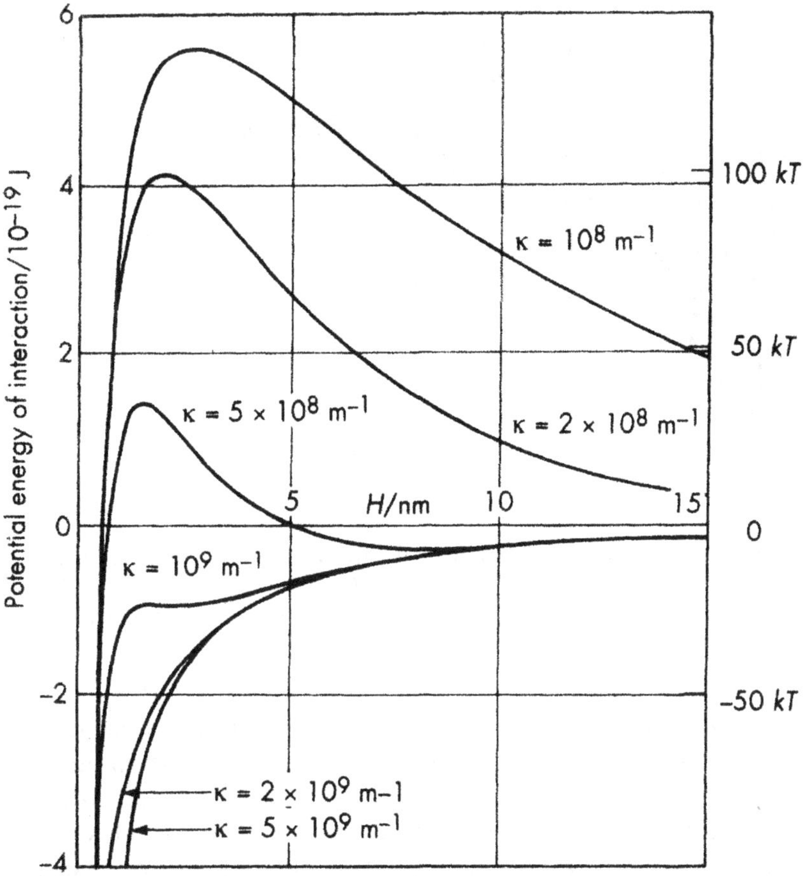

Figure 8.3 The influence of electrolyte concentration, κ, on the total potential energy of interaction of two spherical particles: $a = 10^{-7}$ m; $T = 298$ K; $z = 1$; $A_{11} = 2 \times 10^{-19}$ J; $A_{33} = 0.4 \times 10^{-19}$ J; $\epsilon/\epsilon_0 = 78.5$; $\psi_d = 50$ mV $\approx 2kT/e$. V_R and V_A calculated using equations (8.7), (8.9) and (8.13)

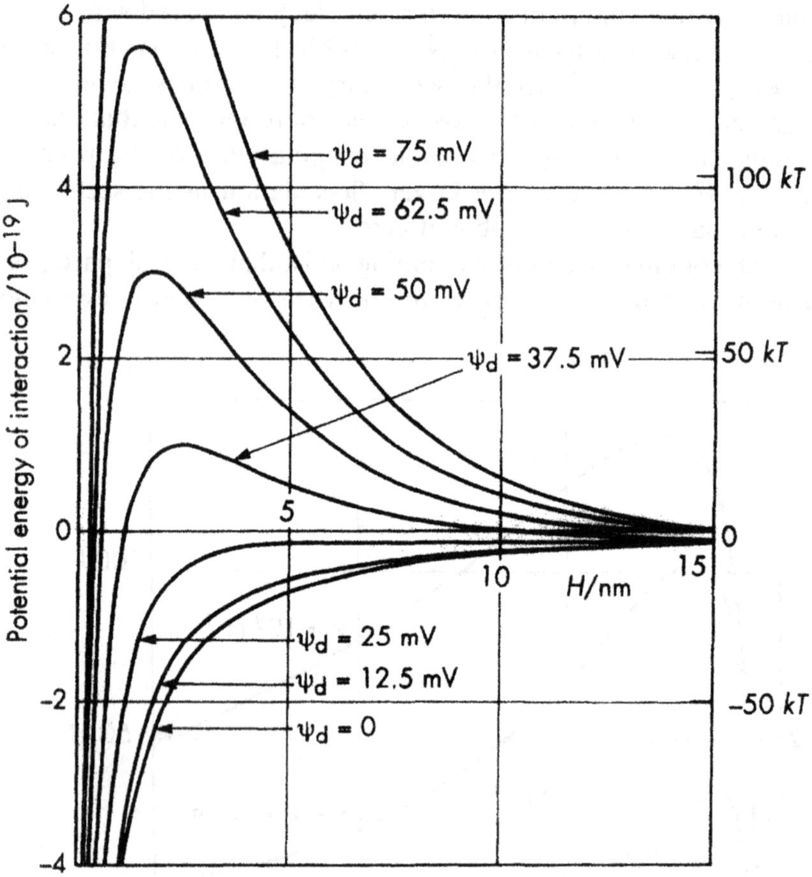

Figure 8.4 The influence of Stern potential, ψ_d, on the total potential energy of interaction of two spherical particles: $a = 10^{-7}$ m; $T = 298$ K; $z = 1$; $A_{11} = 2 \times 10^{-19}$ J; $A_{33} = 0.4 \times 10^{-19}$ J; $\epsilon/\epsilon_0 = 78.5$; $\kappa = 3 \times 10^{8}$ m^{-1}. V_R and V_A calculated using equations (8.7), (8.9) and (8.13)

separation into two phases when the sol is sufficiently concentrated and the electrolyte concentration is too low for coagulation in the primary minimum. One of the phases is a dilute isotropic sol, the other a more concentrated birefringent sol. The particles in the birefringent phase are regularly aligned as parallel rods or plates *c.* 10–100 nm apart (the distance depending on the pH and ionic strength of the sol).

Secondary minimum flocculation is considered to play an important role in the stability of certain emulsions and foams.

Measurement of particle interactions

Owing to their fundamental interest and their practical importance in issues such as colloid stability, much experimental effort has been devoted to the measurement of electric double layer and van der Waals interactions between macroscopic objects at close separations. Such measurements involve balancing the force(s) to be measured with an externally applied force.

Electric double layer interactions

Ottewill and co-workers[106,200] have used a compression method to measure the double-layer repulsion between the plate-like particles of sodium montmorillonite. This is a particularly suitable system for such studies, since the particles are sufficiently thin (*c.* 1 nm) for van der Waals forces to be unimportant and surface roughness is not a problem. The dispersion was confined between a semipermeable filter and an impermeable elastic membrane and an external pressure was applied via a hydraulic fluid so that the volume concentration of particles and, hence, the distance of separation between the particles could be measured as a function of applied pressure.

In another method, Roberts and Tabor[201] measured the electric double layer repulsion between a transparent rubber sphere and a plane glass surface separated by surfactant solution. As the surfaces were brought together, the double-layer interaction caused a distortion of the rubber surface which was monitored interferometrically.

The results of these measurements are in reasonable agreement with predictions based on electric double layer theory, especially at separations greater than $1/\kappa$.

van der Waals interactions

Attractive forces between macroscopic objects have been measured directly by a number of investigators. In the first experiment of this kind, Deryagin and Abricossova[107,202,203] used a sensitive electronic feed-back balance to measure the attraction for a planoconvex polished quartz system from which all residual electric charge had been removed. Relatively large separations were involved and the measured attractive forces were consistent with those predicted by theory, provided that retardation was allowed for.

Recently, measurements in the non-retarded range have been made, the most notable being those of Tabor and co-workers[95,108], on the attraction between cleaved layers of mica stuck to two crossed cylinders. In addition to providing successful tests of the distance dependence of the van der Waals attraction, the effects of adsorbed monolayers have also been studied, again giving reasonable agreement with theoretical predictions.

Structural interactions

Both Hamaker and Lifshitz theories of van der Waals interaction between particles are continuum theories in which the dispersion medium is considered to have uniform properties. At short distances (i.e. up to a few molecular diameters) the discrete molecular nature of the dispersion medium cannot be ignored. In the vicinity of a solid surface, the constraining effect of the solid and the attractive forces between the solid and the molecules of the dispersion medium will cause these molecules to pack, as depicted schematically in Figure 8.5. Moving away from the solid surface, the molecular density will show a damped oscillation about the bulk value. In the presence of a nearby second solid surface, this effect will be even more pronounced. The van der Waals interaction will, consequently, differ from that expected for a continuous dispersion medium. This effect will not be significant at liquid–liquid interfaces where the surface molecules can overlap, and its significance will be difficult to estimate for a rough solid surface.

Israelachvili and co-workers[95] have measured directly the forces between molecularly smooth cleaved mica surfaces separated by organic liquids and have observed a corresponding periodicity of force with separation. The extent to which these short-range interactions (solvation or structural forces) may influence colloid stability and other related phenomena is not entirely clear.

Particle adhesion[105]

The adhesion of colloidal particles to solid substrates is of fundamental and technological importance (e.g. pneumatic transport of powders, printing, filtration, detergency, air pollution). In general,

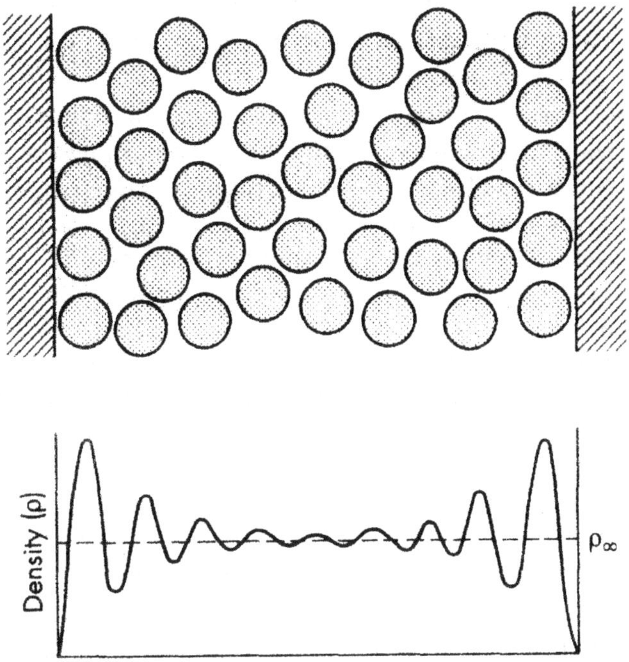

Figure 8.5 Distribution and density of dispersion medium molecules between molecularly smooth solid surfaces

the principles developed in this chapter for particle–particle interactions apply also to particle–solid substrate interactions. Many indirect methods involving centrifugation, vibration, etc., have been developed to measure the force required to detach particles from a solid surface.

Determination and prediction of critical coagulation concentrations

The transition between stability and coagulation, although in principle a gradual one, usually occurs over a reasonably small range of electrolyte concentration, and critical coagulation concentrations can be determined quite sharply. The exact value of the critical coagulation concentration depends upon the criterion which is set for judging whether, or not, the sol is coagulated, and this must remain consistent during a series of investigations.

A common method for measuring critical coagulation concentrations is to prepare a series of about six small test tubes containing equal portions of the sol and to add to each, with stirring, the same volume of electrolyte in concentrations, allowing for dilution, which span the likely coagulation concentration. After standing for a few minutes, an approximate coagulation concentration is noted and a new set of sols is made up with a narrower range of electrolyte concentrations. After standing for a given time (e.g. 2 h), the sols are reagitated (to break the weaker interparticle bonds and bring small particles into contact with larger ones, thus increasing the sharpness between stability and coagulation), left for a further period (e.g. 30 min) and then inspected for signs of coagulation. The critical coagulation concentration can be defined as the minimum electrolyte concentration which is required to produce a visible change in the sol appearance.

An expression for the critical coagulation concentration (c.c.c.) of an indifferent electrolyte can be derived by assuming that a potential energy curve such as $V(2)$ in Figure 8.2 can be taken to represent the transition between stability and coagulation into the primary minimum. For such a curve, the conditions $V = 0$ and $\mathrm{d}V/\mathrm{d}H = 0$ hold for the same value of H. If V_R and V_A are expressed as in equations (8.7) and (8.10), respectively,

$$V = V_\mathrm{R} + V_\mathrm{A} = \frac{32\pi\epsilon a k^2 T^2 \gamma^2}{e^2 z^2} \exp[-\kappa H] - \frac{Aa}{12H} = 0$$

and

$$\frac{\mathrm{d}V}{\mathrm{d}H} = \frac{\mathrm{d}V_\mathrm{R}}{\mathrm{d}H} + \frac{\mathrm{d}V_\mathrm{A}}{\mathrm{d}H} = -\kappa V_\mathrm{R} - \frac{V_\mathrm{A}}{H} = 0$$

from which $\kappa H = 1$; therefore,

$$\frac{32\pi\epsilon a k^2 T^2 \gamma^2}{e^2 z^2} \exp[-1] - \frac{Aa\kappa}{12} = 0$$

giving

$$\kappa_{(\mathrm{coagulation})} = \frac{443.8\epsilon k^2 T^2 \gamma^2}{A e^2 z^2}$$

Substituting $\left(\dfrac{2e^2N_Acz^2}{\epsilon kT}\right)^{1/2}$ for κ (equation (7.6)) gives

$$\text{c.c.c.} = \frac{9.85 \times 10^4 \epsilon^3 k^5 T^5 \gamma^4}{N_A e^6 A^2 z^6} \tag{8.14}$$

For an aqueous dispersion at 25°C, equation (8.14) becomes

$$\text{c.c.c.} = \frac{3.84 \times 10^{-39} \gamma^4}{(A/\text{J})^2 z^6} \text{ mol dm}^{-3} \tag{8.15}$$

A number of features of the Deryagin–Landau–Verwey–Overbeek (D.L.V.O.) theory emerge from these expressions:

1. Since γ limits to unity at high potentials and to $ze\psi_d/4kT$ at low potentials, critical coagulation concentrations are predicted to be proportional to $1/z^6$ at high potentials and to ψ_d^4/z^2 at low potentials (see Figure 8.6). For a typical hydrosol, ψ_d will have an intermediate value in this respect. Taking 75 mV as a typical value for ψ_d (see Figure 7.4), critical coagulation concentrations of inert electrolytes with z = 1, 2 and 3 are, for a given sol, predicted to be in the ratio 100 : 6.7 : 0.8. This is broadly in accord with experimental c.c.c. values, such as those presented in Table 8.1. The experimental values, however, tend to show a significantly stronger dependence on z than predicted above, and this probably reflects increased specific adsorption of counter-ions in the Stern layer with increasing z[109].
2. For a typical experimental hydrosol critical coagulation concentration at 25°C of 0.1 mol dm^{-3} for z = 1, and, again, taking ψ_d = 75 mV, the effective Hamaker constant, A, is calculated to be equal to 8×10^{-20} J. This is consistent with the order of magnitude of A which is predicted from the theory of London–van der Waals forces (see Table 8.3).
3. Critical coagulation concentrations for spherical particles of a given material should be proportional to ϵ^3 and independent of particle size.

The definitions of the term 'critical coagulation concentration' (*a*) in relation to experimental measurements and (*b*) as a means for

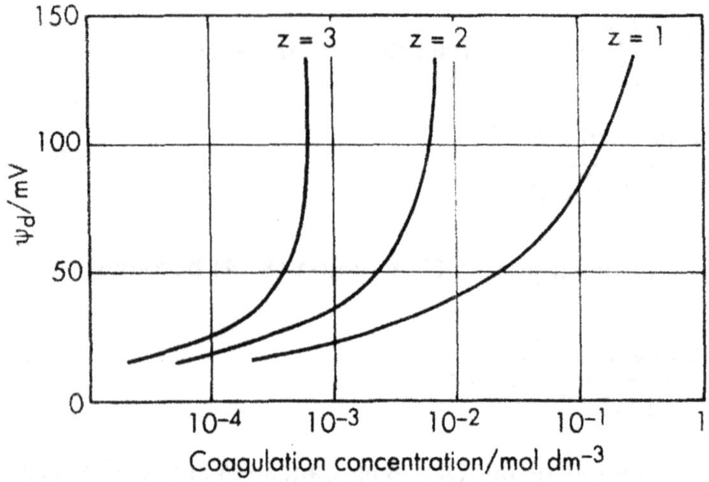

Figure 8.6 Coagulation concentrations calculated from equation (8.15), taking $A = 10^{-19}$ J, for counter-ion charge numbers 1, 2 and 3. The sol is predicted to be stable above and to the left of each curve and coagulated below and to the right

arriving at equation (8.14) are both arbitrary and, no doubt, slightly different from each other. In view of this (in addition to inevitable complications arising from specific ion adsorption and solvation), the results of critical coagulation concentration measurements can only be taken as support for the validity of the D.L.V.O. theory in its broadest outline. To make more detailed tests of stability theories, study of the kinetics of coagulation presents a better line of approach.

Kinetics of coagulation

Lyophobic dispersions are never stable in the thermodynamic sense, but exhibit some degree of instability. From a practical point of view, the word 'stable' is often loosely used to describe a dispersion in which the coagulation rate is slow in relation to its required 'shelf life'.

The rate at which a sol coagulates depends on the frequency with which the particles encounter one another and the probability that their thermal energy is sufficient to overcome the repulsive potential energy barrier to coagulation when these encounters take place.

The rate at which particles aggregate is given by

$$-\frac{dn}{dt} = k_2 n^2$$

where n is the number of particles per unit volume of sol at time t, and k_2 is a second-order rate constant.

Integrating, and putting $n = n_0$ at $t = 0$, gives

$$\frac{1}{n} - \frac{1}{n_0} = k_2 t \tag{8.16}$$

During the course of coagulation k_2 usually decreases, and sometimes an equilibrium state is reached with the sol only partially coagulated. This may be a consequence of the height of the repulsion energy barrier increasing with increasing particle size. In experimental tests of stability theories it is usual to restrict measurements to the early stages of coagulation (where the aggregating mechanism is most straightforward), using moderately dilute sols.

The particle concentration during early stages of coagulation can be determined directly, by visual particle counting, or indirectly, from turbidity (spectrophotometric or light scattering) measurements[23,110,204]. If necessary, coagulation in an aliquot of sol can be halted prior to examination by the addition of a small amount of a stabilising agent, such as gelatin. The rate constant k_2 is given as the slope of a plot of $1/n$ against t.

In most colloid stability studies, coagulation rates are measured, as far as possible, under *perikinetic* (non-agitated) conditions, where particle–particle encounters are solely the result of Brownian motion. Particle aggregation under *orthokinetic* (agitated) conditions is of technological importance. Agitation increases the particle flux by a factor which depends on the third power of the collision diameter of the particles. With large particles, such as in emulsions, orthokinetic aggregation can occur at up to as much as 10^4 times the perikinetic rate; but, with particles at the lower end of the colloidal size range, stirring has relatively little effect on their rate of aggregation.

The potential energy barrier to coagulation can be reduced to zero by the addition of excess electrolyte, which creates a situation in which every encounter between the particles leads to permanent contact. The theory of rapid (diffusion-controlled) coagulation was developed by Smoluchowski[110]. For a monodispersed sol containing spherical particles, and considering only the aggregation of single particles to form doublets

$$n = \frac{n_0}{(1 + 8\pi D a n_0 t)} \tag{8.17}$$

where a is the effective radius of the particles and D is the diffusion coefficient. Substituting $D = kT/6\pi\eta a$ (equation (2.6)) and combining equations (8.16) and (8.17) gives

$$k_2^0 = \frac{4kT}{3\eta} \tag{8.18}$$

where k_2^0 is the rate constant for diffusion-controlled coagulation.

For a hydrosol at room temperature, the time $t_{1/2}$ in which the number of particles is halved by diffusion-controlled coagulation is calculated from the above equations to be of the order of $10^{11}/n_0$ seconds, if n_0 is expressed in the unit, particles cm^{-3}. In a typical dilute hydrosol, the number of particles per cm^3 may be about 10^{10}–10^{11}, and so, on this basis, $t_{1/2}$ should be of the order of a few seconds.

Rapid coagulation is, in fact, not quite as simple as this, because the last part of the approach of two particles is (*a*) slowed down because it is difficult for liquid to flow away from the narrow gap between the particles, and (*b*) accelerated by the van de Waals attraction between the particles. Lichtenbelt and co-workers[205] have measured rapid coagulation rates by a stopped-flow method and found them, typically, to be about half the rate predicted according to equation (8.18).

When there is a repulsive energy barrier, only a fraction $1/W$ of the encounters between particles lead to permanent contact. W is known as the stability ratio – i.e.

$$W = \frac{k_2^0}{k_2} \tag{8.19}$$

A theoretical expression relating the stability ratio to the potential energy of interaction has been derived by Fuchs[110]:

$$W = 2a \int_{2a}^{\infty} \frac{\exp[V/kT]}{R^2} \, dR \tag{8.20}$$

Theoretical relationships between the stability ratio and electrolyte concentration can be obtained by numerical solution of this integral for given values of A and ψ_d. Figure 8.7 shows the results of

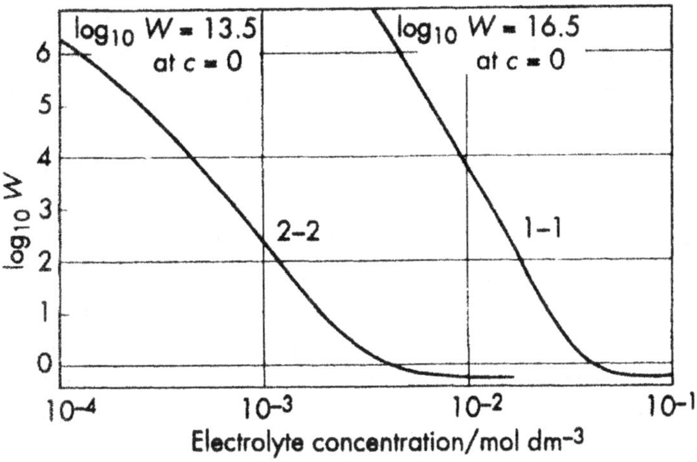

Figure 8.7 Theoretical dependence of stability ratio on electrolyte concentration calculated from equation (8.2) for $a = 10^{-8}$ m, $A = 2 \times 10^{-19}$ J and $\psi_d = 76.8$ mV $= 3kT/e$. At high electrolyte concentrations $W < 1$ owing to coagulation being accelerated by van der Waals attractive forces (reduced flow rate in the narrow inter-particle gap has not been allowed for) (*By courtesy of Elsevier Publishing Company*)

calculations for 1–1 and 2–2 electrolytes. For constant ψ_d, a linear relationship between log *W* and log *c* is predicted for practically the whole of the slow coagulation region.

An alternative approach (which is more convenient, but more approximate) is that of Reerink and Overbeek[196], who have combined an approximate form of equation (8.20),

$$W \approx \frac{1}{2\kappa a}\exp\left[\frac{V_{max}}{kT}\right]$$

with equations (8.7) and (8.10) to derive a theoretical expression which also predicts a linear relationship between log *W* and log *c* at constant ψ_d. For a temperature of 25°C and with the particle radius expressed in metres, the resulting equation takes the form

$$\log W = \text{constant} - 2.06\times10^{9}\left(\frac{a\gamma^2}{z^2}\right)\log c \tag{8.21}$$

According to this approximation, d log *W*/d log *c* for the example of $a = 10^{-8}$ m and $\psi_d = 3kT/e$ chosen in Figure 8.7 is equal to 9 for 1–1 electrolytes and 4.8 for 2–2 electrolytes, whereas the more exact calculations via equation (8.20) give slopes of 7 and 4.5, respectively[96].

Coagulation rates have been measured as a function of electrolyte concentration for a number of sols[96,196,204,206], and the predicted linear relationship between log *W* and log *c* in the slow-coagulation region seems to be well confirmed. In addition, the experimental values of d log *W*/d log *c*, although somewhat variable, are of the right order of magnitude compared with theoretical slopes.

Figure 8.8 shows some interesting results which have been obtained by Fairhurst and Smith[206] for the coagulation of silver iodide hydrosols at various pI values. As the pI is decreased (and the potential ψ_0 becomes more negative) the slope d log *W*/d log *c* and the critical coagulation concentration (which is the concentration which corresponds to an arbitrarily chosen low value of *W*) increase, as expected, until a pI of about 6 is reached. However, as the pI is reduced below 6, d log *W*/d log *c* and the critical coagulation concentration decrease. This apparently anomalous observation (and the corresponding maximum in the zeta potential curve – Figure 7.4) may be a consequence of the discreteness of charge effect, described on page 188.

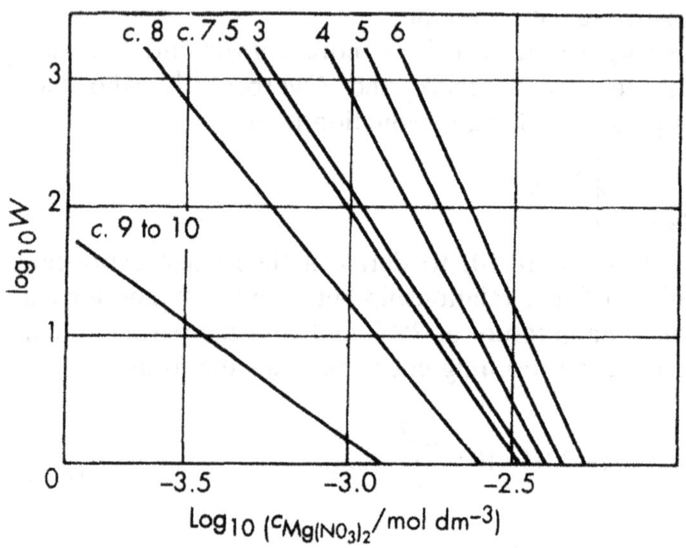

Figure 8.8 Plots of log *W* versus log *c* for coagulation of AgI sols at various pI values by magnesium nitrate[206] (*By courtesy of Dr D. Fairhurst and Dr A.L. Smith*)

Experimental data are generally not in accord with the theoretical prediction in equation (8.21) regarding particle size[96,196,204]. For example, Ottewill and Shaw[204] found no systematic variation in d log W/d log c for a number of monodispersed carboxylated polystyrene latex dispersions with the particle radius ranging from 30 nm to 200 nm. This problem still remains unresolved.

Peptisation

Peptisation is a process in which dispersion is achieved (with little or no agitation) by changing the composition of the dispersion medium. Methods of peptisation include addition of polyvalent co-ions (e.g. polyphosphate ions to a negatively charged coagulated dispersion), addition of surfactants, dilution of the dispersion medium and dialysis. In each case, V_R is modified so as to create a potential energy maximum (see Figures 8.2–8.4) to act as a barrier against recoagulation.

Simple consideration of lyophobic sols in terms of their potential energy of interaction curves does not lead one to expect peptisation, since any energy barrier to coagulation would involve on the other side an even greater energy barrier to peptisation. Nevertheless, peptisation of lyophobic sols is possible, especially when only a short time is allowed to elapse between coagulation and peptisation.

There are several possible explanations of this phenomenon[99].

1. Both specific adsorption (particularly, solvation) at the particle surfaces and the difficulty with which dispersion medium flows from the narrow gap between the particles may hinder particle approach to the small separation which corresponds to the primary minimum.
2. Owing to the time lag in the adjustment of potential and/or charge as particles approach or move away from one another (see page 212), coagulation and peptisation will occur at potentials which are, respectively, lower than and greater than the equilibrium potential.

Sedimentation volume and gelation

As the particles of a dispersion usually have a density somewhat different from that of the dispersion medium, they will tend to accumulate under the influence of gravity at the bottom or at the surface. A sedimentation velocity (see Table 2.2) of up to *c.* 10^{-8} m s^{-1} is usually counteracted by the mixing tendencies of diffusion and convection. Particle aggregation, of course, enhances sedimentation.

When sedimentation does take place, the volume of the final sediment depends upon the extent of aggregation. Relatively large peptised particles pack efficiently, to give a dense sediment which is difficult to redisperse, whereas aggregated particles bridge readily and give a loose sediment which (provided that the particles are not held together too strongly) is more easily dispersed (see Figure 8.9). In extreme cases, the sedimentation volume may equal the whole volume, and this can lead to the paradoxical situation where a small amount of aggregating agent produces a sediment while a larger amount does not. Gentle stirring usually reduces the sedimentation volume.

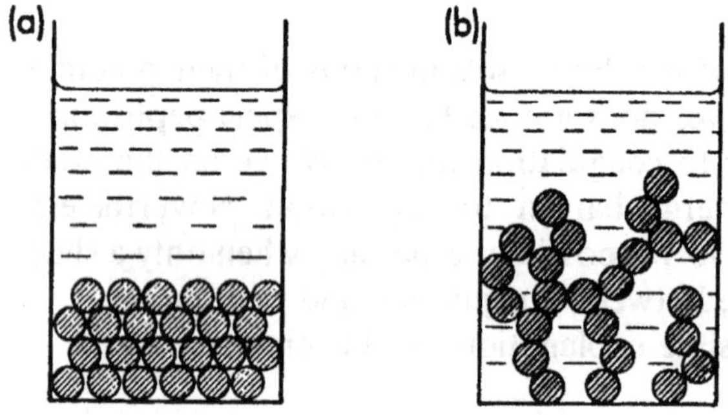

Figure 8.9 Sedimentation volumes for (a) peptised and (b) aggregated particles

When the particles aggregate to form a continuous network structure which extends throughout the available volume and immobilises the dispersion medium, the resulting semi-solid system is called a *gel*. The rigidity of a gel depends on the number and the strength of the interparticle links in this continuous structure.

Systems containing lyophilic material

Polymer solutions

Macromolecular solutions are stabilised by a combination of electric double layer interaction and solvation, and both of these stabilising influences must be weakened sufficiently before precipitation will take place. For example, gelatin has a sufficiently strong affinity for water to be soluble (unless the electrolyte concentration is very high) even at its isoelectric pH, where there is no double-layer interaction. Casein, on the other hand, exhibits weaker hydrophilic behaviour and is precipitated from aqueous solution when the pH is near to the isoelectric point.

Owing to their affinity for water, hydrophilic colloids are unaffected by the small amounts of added electrolyte which cause hydrophobic sols to coagulate but are often precipitated (salted out) when the electrolyte concentration is high. The ions of the added electrolyte dehydrate the hydrophilic colloid by competing for its water of hydration. The salting-out efficacy of an electrolyte, therefore, depends upon the tendencies of its ions to become hydrated. Thus, cations and anions can be arranged in the following lyotropic series of approximately decreasing salting-out power:

$$Mg^{2+} > Ca^{2+} > Sr^{2+} > Ba^{2+}$$

$$> Li^{+} > Na^{+} > K^{+} > NH_4^{+} > Rb^{+} > Cs^{+}$$

and

$$\text{citrate}^{3-} > SO_4^{2-} > Cl^{-} > NO_3^{-} > I^{-} > CNS^{-}$$

Ammonium sulphate, which has a high solubility, is often used to precipitate proteins from aqueous solution.

Lyophilic colloids can also be desolvated (and precipitated if the electric double layer interaction is sufficiently small) by the addition of non-electrolytes, such as acetone or alcohol to aqueous gelatin solution and petrol ether to a solution of rubber in benzene.

Dispersions containing stabilising agents[111–114]

The stability of lyophobic sols can often be enhanced by the addition of soluble lyophilic material which adsorbs on to the particle surfaces. Such adsorbed material is sometimes called a *protective agent*. The stabilisation mechanism is usually complex and a number of factors may be involved.

Lyophilic stabilisation is particularly important in non-aqueous systems[115], e.g. oil-based paints, and in systems of very high particle concentration where electrostatic stabilisation is of limited effectiveness. It is also essential in biological systems, e.g. blood, where the electrolyte concentration is often sufficiently high to render electrostatic stabilisation ineffective.

Effect on electric double layer interactions

If the stabilising agent is ionised and carries a charge of the same sign as that on the particles (e.g. anionic surfactant adsorbed on negatively charged particles), then electric double layer repulsion will be enhanced. The adsorbed stabilising agent (even if non-ionic) will influence electrostatic interactions by causing a displacement of the Stern plane away from the particle surface; this will increase the range of electric double layer repulsion and, as such, enhance stability.

Effect on van der Waals interactions

Adsorbed layers of stabilising agent may cause a significant lowering of the effective Hamaker constant and, therefore, a weakening of the interparticle van der Waals attraction. This effect has been considered by Vold[207] and by Vincent and co-workers[208] in terms of the Hamaker microscopic treatment of dispersion forces.

For the interaction depicted in Figure 8.10,

$$-12V_A = F_{11}\,(A_{11}^{1/2} - A_{22}^{1/2})^2 + F_{22}(A_{22}^{1/2} - A_{33}^{1/2})^2$$
$$+2F_{12}(A_{11}^{1/2} - A_{22}^{1/2})(A_{22}^{1/2} - A_{33}^{1/2})$$

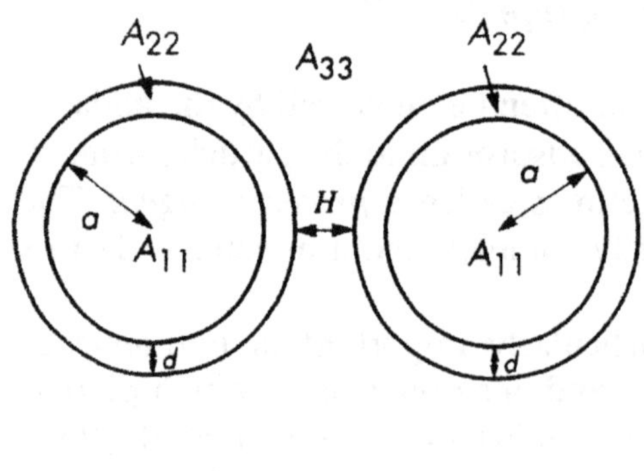

Figure 8.10

where

$$F = \frac{y}{x^2 + xy + x} + \frac{y}{x^2 + xy + x + y} + 2\ln\left(\frac{x^2 + xy + x}{x^2 + xy + x + y}\right)$$

with x and y defined as follows:

$$\begin{aligned} &\text{for } F_{12}, \quad x = (H + 2d)/2a; && y = 1 \\ &\text{for } F_{22}, \quad x = H/2(a + d); && y = 1 \\ &\text{for } F_{12}, \quad x = (H + d)/(2a + d); && y = (a + d)/a \end{aligned} \tag{8.22}$$

Langbein[209] has derived a corresponding expression for this so-called 'Vold effect' which is based on the Lifshitz macroscopic treatment of dispersion forces.

Steric stabilisation

The stability of many 'protected' colloidal dispersions cannot be explained solely on the basis of electric double layer repulsion and van der Waals attraction; other stabilising mechanisms must be investigated. 'Steric stabilisation' is a name which is used (somewhat loosely) to describe several different possible stabilising mechanisms involving adsorbed macromolecules. These include the following:

1. An encounter between particles could involve desorption of stabilising agent at the point of contact. Since adsorption is a spontaneous process, $\Delta G_{\text{ads.}}$ is negative and $\Delta G_{\text{des.}}$ positive. This positive free energy of desorption corresponds to particle–particle repulsion and enhanced stability. However, for polymeric material, adsorption and desorption processes tend to be slow compared with the time of a typical particle–particle encounter[99] and so, in such cases, the attainment of a 'primary minimum' coagulated condition is unlikely.
2. When particles collide, their adsorbed layers may be compressed without penetrating into one another. This 'denting' mechanism will reduce the configurations available to the adsorbed polymer molecules; therefore, there will be a decrease in entropy and an increase in free energy, and stability will be enhanced by an elastic

effect. However, it is likely that this mechanism is not significant in practice.

3. The adsorbed layers between the particles may interpenetrate and so give a local increase in the concentration of polymer segments. Depending on the balance between polymer–polymer and polymer–dispersion medium interactions, this may lead to either repulsion or attraction by an osmotic mechanism. Enthalpic and entropic changes will be involved. If interpenetration takes place to a significant extent, elastic repulsion will also operate.

Steric stabilisers are usually block copolymer molecules (e.g. poly (ethylene oxide) surfactants), with a lyophobic part (the 'anchor' group) which attaches strongly to the particle surface, and a lyophilic chain which trails freely in the dispersion medium. The conditions for stabilisation are similar to those for polymer solubility outlined in the previous section. If the dispersion medium is a good solvent for the lyophilic moieties of the adsorbed polymer, interpenetration is not favoured and interparticle repulsion results; but if, on the other hand, the dispersion medium is a poor solvent, interpenetration of the polymer chains is favoured and attraction results. In the latter case, the polymer chains will interpenetrate to the point where further interpenetration is prevented by elastic repulsion.

The free energy change which takes place when polymer chains interpenetrate is influenced by factors such as temperature, pressure and solvent composition. The point at which this free energy change is equal to zero is known as the θ (theta)-point and such a solvent is called a θ-solvent. More formally, a θ-point is defined as one where the second virial coefficient of the polymer chains is equal to zero. It can be determined by light scattering and by osmometry.

The positive ΔG for polymer chain interpenetration which leads to steric stabilisation is given in terms of the corresponding enthalpy and entropy changes by $\Delta G = \Delta H - T\Delta S$, and therefore, stabilisation could be the result of a positive ΔH and/or a negative ΔS. A positive ΔH would reflect the release of bound solvent from the polymer chains as they interpenetrate and a negative ΔS would reflect loss of configurational freedom as the polymer chains interpenetrate. If ΔH is positive and ΔS negative, the dispersion will be sterically stabilised at all accessible temperatures; however, if ΔH and ΔS are both positive, the dispersion should flocculate on heating above the θ-temperature (enthalpic stabilisation), whereas if ΔH and ΔS are both negative, the dispersion should flocculate on cooling below the θ-temperature (entropic stabilisation).

These effects have been observed for both aqueous and non-aqueous media and good correlation between the point of incipient flocculation and the θ-temperature is well established[112]. The transition from stability to instability usually occurs over a very narrow temperature range (1 or 2 K). Enthalpic stabilisation tends to be the more common in aqueous media and entropic stabilisation the more common in non-aqueous media. Owing to the elastic effect, aggregation into a deep primary minimum does not take place (as is possible with lyophobic sols) and redispersion takes place readily on reverting to better than θ-solvent conditions.

Table 8.4 Classification of sterically stabilised dispersions and comparison of critical flocculation temperatures (c.f.t) with theta-temperatures[112] (*By courtesy of Academic Press Inc.*)

Stabiliser	$M_r/10^3$	*Dispersion medium*	*Classification*	*c.f.t./K*	θ/K
Poly (ethylene oxide)	10 96 1000	0.39 mol dm^{-3} $MgSO_4$(aq.)	enthalpic	318 ± 2 316 ± 2 317 ± 2	315 ± 3
Poly (acrylic acid)	9.8 51.9 89.7	0.2 mol dm^{-3} HCl(aq.)	entropic	287 ± 2 283 ± 2 281 ± 1	287 ± 5
Polyiso-butylene	23 150	2-methyl butane	enthalpic	325 ± 1 325 ± 1	325 ± 2

Several quantitative theories of steric stabilisation have been developed over the last few decades[112,114,210–212].

The forces between sterically stabilised particles have been measured with a compression cell (see page 223)[213], and have been shown (as expected in the light of the foregoing discussion) to be short-range, with a range comparable with twice the contour length of the lyophilic chains. For sterically stabilised systems the total interaction energy can be written as

$$V = V_R + V_A + V_S \tag{8.23}$$

and potential energy diagrams will be as schematically depicted in Figure 8.11, with entry into a deep primary minimum made virtually impossible by the steric interactions.

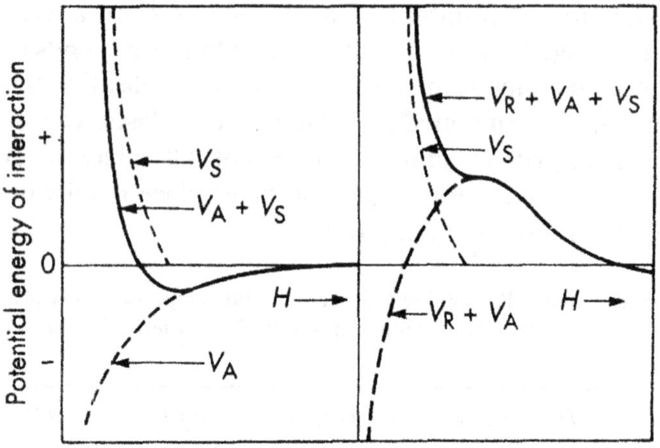

Figure 8.11 Schematic interaction energy diagrams for sterically stabilised particles: (a) in the absence of electric double layer repulsion ($V = V_A + V_S$), (b) with electric double layer repulsion ($V = V_R + V_A + V_S$)

Sensitisation

In certain cases, colloidal dispersions are made more sensitive to aggregation by the addition of small quantities of materials which, if used in larger amounts, would act as stabilising agents. Several factors may contribute to such observations:

1. If the sol particles and the additive are oppositely charged, sensitisation results when the concentration (and adsorption) of the additive is such that the charge on the particles is neutralised, whereas stabilisation results at higher concentrations because of a reversal of the charge and increasing steric effects.
2. At low concentrations, surface-active additives may form a first adsorbed layer on the sol particles with the lyophobic part orientated outwards, thus sensitising the sol. At higher concentrations a second, oppositely orientated, layer would then give protection[214].

3. Long-chain additives, such as gelatin, can sometimes bring about a rather loose flocculation by a bridging mechanism in which the molecules are adsorbed with part of their length on two or more particles[111,215–216]. Such flocculation normally occurs over a narrow range of additive concentrations; at high concentrations protective action is obtained, since bridging can occur only through particle collisions under conditions where further adsorption of the additive is possible.

Dispersions of hydrophilic particles in oil media can be flocculated by traces of water, which form thin interconnecting films between the particles[116].

Another interesting phenomenon is that of depletion flocculation. This can be observed with dispersions (e.g. lattices) which contain inert additives, such as free polymer, non-ionic surfactant or even small (e.g. silica) particles. As the latex particles approach one another, the gaps between them become too small to accommodate the above additives, but the kinetic energy of the particles may be sufficient to enable them to be expelled from the gap; i.e. a 'de-mix' occurs, for which ΔG is positive. When this 'de-mix' has been achieved, an osmotic situation exists in which the remaining pure dispersion medium will tend to flow out from the gap between the particles in order to dilute the bulk dispersion medium, thus causing the particles to flocculate.

Stability control

Particle aggregation and sedimentation volume are important in many practical situations, as illustrated by the following examples.

1. *Agricultural soil.* It is necessary to maintain agricultural soil in a reasonably aggregated state in order to achieve good aeration and drainage, and treatment with coagulants, such as calcium salts (lime or gypsum) or organic polyelectrolytes (so-called soil conditioners), is common practice. An extreme example of the effect of soil peptisation occurs when agricultural land is flooded with sea-water. The calcium ions of the naturally occurring clay minerals in the soil exchange with the sodium ions of the sea-water. Subsequent leaching of the sodium ions from the soil by

rain-water leads to peptisation and the soil packs into a hard mass which is unsuitable for plant growth. Conversely, water seepage from reservoirs can be reduced by initial flooding with sea-water.

2. *Oil well drilling*. In the drilling of oil wells a drilling mud (usually a bentonite clay suspension) is used (*a*) as a coolant, (*b*) for removing the cuttings from the bore-hole and (*c*) to seal the sides of the bore-hole with an impermeable filter cake. The pumping and sealing features of this operation are most effective if the drilling mud is peptised; however, a certain amount of mud rigidity is required to reduce sedimentation of the cuttings, especially during an interruption of circulation. These opposite requirements are somewhat reconciled by maintaining the drilling mud in a partially coagulated, thixotropic (page 254) state. If the drilling mud stiffens, partial redispersion can be effected by the addition of a small amount of a peptising agent, such as a polyphosphate. The plate-like particles of clays often have negatively charged faces and positively charged edges when in contact with aqueous media, and aggregate quite readily by an edge-to-face mechanism to form a gel structure, even at moderately low clay concentrations[18]. The main function of the polyphosphate is to reverse the positive charge on the edges of the clay particles. The relatively small edge area makes this process economically attractive.
3. *Sewage treatment and water purification*. Industrial waste water and domestic sewage contains a variety of particulate matter and surfactant (mostly anionic). The zeta potentials of the particles are usually in the range -10 to -40 mV. Considerable purification can be effected by the addition of small amounts of sodium hydrogen carbonate plus aluminium sulphate and agitating the mix. The aluminium ions are hydrolysed to give a polymeric hydrous oxide gel network in which the suspended particles become entrapped and bound together by a bridging mechanism. The pH is adjusted to near pH 6 to give a slightly positive zeta potential (*c.* $+5$ mV). At this zeta potential, electrostatic stabilisation is insignificant, but a significant removal of anionic surfactant by adsorption on to the positively charged flocs takes place. In the final stage of water purification, most of the remaining particulate matter can be removed by the addition of a few parts per million of high molecular mass polyacrylamide, again by a bridging mechanism.

4. *Paints*. The particles in pigmented paints are often large enough to settle even when peptised; therefore, it is desirable that they should be aggregated to a certain extent to facilitate redispersion.

Some other practical situations where particle aggregation is important include the precipitation of colloidal mud at the mouth of a river due to the salinity of the sea-water exceeding the critical coagulation concentration, land (e.g. mountainside) stability, building and road foundations, the retention of a porous structure in filtration, mineral processing[117] and paper making. Control of particle aggregation is also of primary importance in adhesives, inks, pharmaceuticals, cosmetics, foodstuffs and lubricants.

9 Rheology

Introduction

Rheology is the science of the deformation and flow of matter, and its study has contributed much towards clarifying ideas concerning the nature of colloidal systems. It is a subject of tremendous and increasing technological importance – in many industries, such as rubber, plastics, food, paint and textiles, the suitability of the products involved is to a large extent judged in terms of their mechanical properties[118]. In biology and medicine (particularly haematology) rheological behaviour is also of major importance[119].

The most straightforward rheological behaviour is exhibited on the one hand by Newtonian viscous fluids and on the other by Hookean elastic solids. However, most materials, particularly those of a colloidal nature, exhibit mechanical behaviour which is intermediate between these two extremes, with both viscous and elastic characteristics in evidence. Such materials are termed *viscoelastic*.

There are two general approaches to rheology, the first being to set up mathematical expressions which describe rheological phenomena without undue reference to their causes, and the second, with which the following discussion is mainly concerned, is to correlate observed mechanical behaviour with the detailed structure of the material in question. This is not an easy task. The rheological behaviour of colloidal dispersions depends mainly on the following factors:

1. Viscosity of the dispersion medium.
2. Particle concentration.
3. Particle size and shape.
4. Particle–particle and particle–dispersion medium interactions.

Because of the complications involved, this aspect of rheology is still in many respects a mainly descriptive science. However, in recent years considerable advances have been made towards understanding rheological behaviour and putting it on to a quantitative basis[120].

For convenience, this chapter has been divided into three sections in which the viscosity of dilute solutions and dispersions, non-Newtonian flow, and the viscoelastic properties of semi-solid systems are discussed.

Viscosity

Newtonian viscosity

The viscosity of a liquid is a measure of the internal resistance offered to the relative motion of different parts of the liquid. Viscosity is described as Newtonian when the shearing force per unit area σ between two parallel planes of liquid in relative motion is proportional to the velocity gradient D between the planes – i.e.

$$\sigma = \eta D \tag{9.1}$$

where η is the *coefficient of viscosity*. The dimension of η is, therefore, (mass) $(\text{length})^{-1}$ $(\text{time})^{-1}$.

For most pure liquids and for many solutions and dispersions, η is a well-defined quantity for a given temperature and pressure which is independent of σ and D, provided that the flow is streamlined (i.e. laminar). For many other solutions and dispersions, especially if concentrated and if the particles are asymmetric and/or aggregated deviations from Newtonian flow are observed. The main causes of non-Newtonian flow are the formation of a structure throughout the system and orientation of asymmetric particles caused by the velocity gradient.

Measurement of viscosity[121]

Capillary flow methods

The most frequently employed methods for measuring viscosities are based on flow through a capillary tube. The pressure under which the liquid flows furnishes the shearing stress.

The relative viscosities of two liquids can be determined by using a

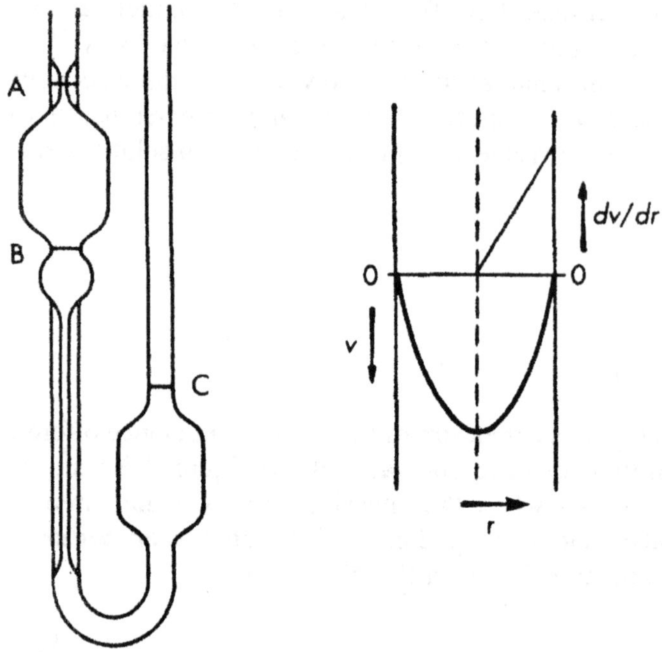

Figure 9.1 An Ostwald viscometer

simple Ostwald viscometer (Figure 9.1). Sufficient liquid is introduced into the viscometer for the levels to be at B and C. Liquid is then drawn up into the left-hand limb until the liquid levels are above A and at the bottom of the right-hand bulb. The liquid is then released and the time for the left-hand meniscus to pass between the marks A and B is measured.

Since the pressure at any instant driving the liquid through the capillary is proportional to its density,

$$\eta = k\rho t \tag{9.2}$$

where k is the viscometer constant, ρ is the density of the liquid and t is the flow time. Therefore, for two different liquids,

$$\frac{\eta_1}{\eta_2} = \frac{\rho_1 t_1}{\rho_2 t_2} \tag{9.3}$$

Accurate thermostatting is necessary, owing to the marked dependence of viscosities on temperature. Dust and fibrous materials, which might block the capillary, must be removed from the liquid prior to its introduction into the viscometer. A viscometer is selected which gives a flow time in excess of *c*. 100 s; otherwise a kinetic energy correction is necessary.

The capillary method is simple to operate and precise (*c*. 0.01–0.1 per cent) in its results, but suffers from the disadvantage that the rate of shear varies from zero at the centre of the capillary to a maximum (which decreases throughout the determination) at the wall. Thus, with asymmetric particles a viscosity determination in an Ostwald viscometer could cover various states of orientation and the measured viscosity, although reproducible, would have little theoretical significance.

Rotational methods

Concentric cylinder and cone and plate instruments are particularly useful for studying the flow behaviour of non-Newtonian liquids.

In the first of these techniques an approximation to uniform rate of shear throughout the sample is achieved by shearing a thin film of the liquid between concentric cylinders. The outer cylinder can be rotated (or oscillated) at a constant rate and the shear stress measured in terms of the deflection of the inner cylinder, which is suspended by a torsion wire (Figure 9.2); or the inner cylinder can be rotated (or oscillated) with the outer cylinder stationary and the resistance offered to the motor measured.

If ω is the angular velocity of the outer cylinder and θ the angular deflection of the inner cylinder, the coefficient of viscosity of the liquid is given by

$$\eta = \frac{k\theta}{\omega} \tag{9.4}$$

where k is an apparatus constant (usually obtained by calibration with a liquid of known viscosity). The necessity for an end-correction can be avoided if the inner cylinder is appropriately cone-shaped at its end, or if air is entrapped in a hollowed-out inner cylinder base.

Cone and plate instruments (see Figure 9.3) permit the velocity gradient to be kept constant throughout the sample, and are particularly useful for studying highly viscous materials. A very versatile cone and plate rheometer, known as a rheogoniometer, has been developed by Weissenberg, which enables both tangential

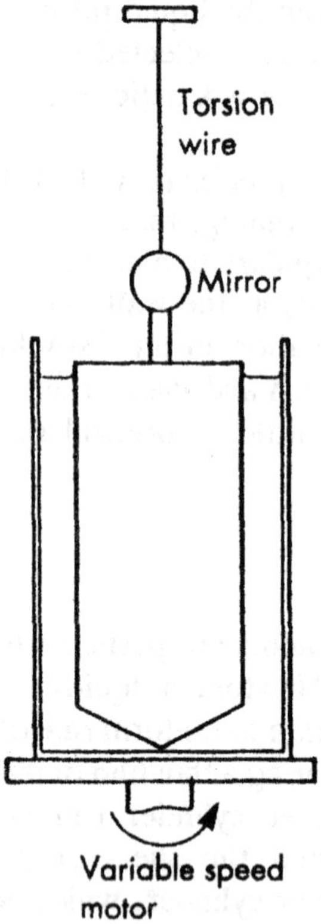

Figure 9.2 A concentric cylinder (Couette) viscometer

forces and normal forces (i.e. forces which tend to push the cone upwards; see page 261) to be measured either in rotation or in oscillation.

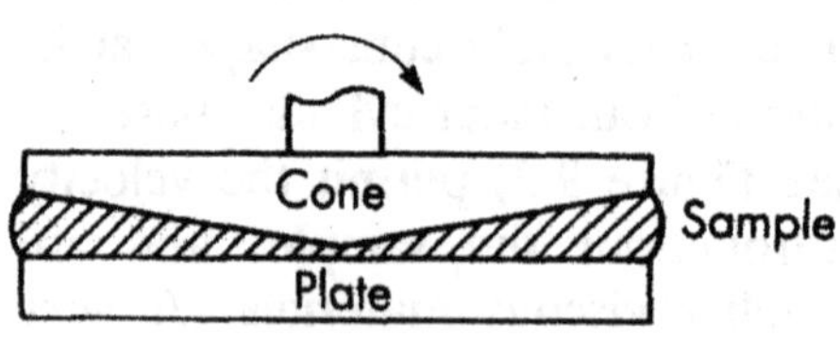

Figure 9.3 Cone and plate method

Viscosities of dilute colloidal solutions and dispersions

Functions of viscosity

When colloidal particles are dispersed in a liquid, the flow of the liquid is disturbed and the viscosity is higher than that of the pure liquid. The problem of relating the viscosities of colloidal dispersions (especially when dilute) with the nature of the dispersed particles has been the subject of much experimental investigation and theoretical consideration. In this respect, viscosity increments are of greater significance than absolute viscosities, and the following functions of viscosity are defined:

η_0 = viscosity of pure solvent or dispersion medium

η = viscosity of solution or dispersion

η/η_0 = relative viscosity (or viscosity ratio)

$\eta_i = \eta/\eta_0 - 1$ = relative viscosity increment (or viscosity ratio increment)

η_i/c = reduced viscosity (or viscosity number)

$$[\eta] = \lim_{c\to 0} \frac{\eta_i}{c} = \lim_{c\to 0} \frac{\eta}{\eta_0}$$

= intrinsic viscosity (or limiting viscosity number)

From the above expressions it can be seen that reduced and intrinsic viscosities have the unit of reciprocal concentration. When one considers particle shape and solvation, however, concentration is generally expressed in terms of the volume fraction ϕ of the particles (i.e. volume of particles/total volume) and the corresponding reduced and intrinsic viscosities are, therefore, dimensionless.

Spherical particles

Einstein (1906) made a hydrodynamic calculation (under assumptions similar to those of Stokes; see page 22) relating to the disturbance of the flow lines when identical, non-interacting, rigid, spherical particles are dispersed in a liquid medium, and arrived at the expression

$$\eta = \eta_0 (1 + k\phi)$$

where k is a constant equal to 2.5 – i.e.

$$\eta_i = 2.5\phi \quad \text{or} \quad [\eta]_\phi = 2.5 \qquad (9.5)$$

The effect of such particles on the viscosity of a dispersion depends, therefore, only on the total volume which they occupy and is independent of their size.

The validity of Einstein's equation has been confirmed experimentally for dilute suspensions ($\phi < c\,0.02$) of glass spheres, certain spores and fungi, polystyrene particles, etc., in the presence of sufficient electrolyte to eliminate charge effects.

For dispersions of non-rigid spheres (e.g. emulsions) the flow lines may be partially transmitted through the suspended particles, making k in Einstein's equation less than 2.5.

The non-applicability of the Einstein equation at moderate concentrations is mainly due to an overlapping of the disturbed regions of flow around the particles. A number of equations, mostly of the type $\eta = \eta_0 (1 + a\phi + b\phi^2 + \ldots)$, have been proposed to allow for this.

Solvation and asymmetry

The volume fraction term ϕ in viscosity equations must include any solvent which acts kinetically as a part of the particles. The intrinsic viscosity is, therefore, proportional to the solvation factor (i.e. the ratio of solvated and unsolvated volumes of dispersed phase). The, solvation factor will usually increase with decreasing particle size.

Particle asymmetry has a marked effect on viscosity and a number of complex expressions relating intrinsic viscosity (usually extrapolated to zero velocity gradient to eliminate the effect of orientation) to axial ratio for rods, ellipsoids, flexible chains, etc., have been proposed. For randomly orientated, rigid, elongated particles, the intrinsic viscosity is approximately proportional to the square of the axial ratio.

Electroviscous effects

When dispersions containing charged particles are sheared, extra energy is required to overcome the interaction between ions in the double layers around the particles and the electrical charge on the particle surfaces, thus leading to an increased viscosity.

For charged flexible chains, in addition to the above effect (which is usually small), the nature of the double layer influences the chain configuration. At low ionic strengths the double-layer repulsions between the various parts of the flexible chain have a relatively long range and tend to give the chain an extended configuration, whereas at high ionic strengths the range of the double-layer interactions is less, which permits a more coiled configuration. Therefore, the viscosity decreases with increasing ionic strength, often in a marked fashion.

Polymer relative molecular masses from viscosity measurements

Viscosity measurements cannot be used to distinguish between particles of different size but of the same shape and degree of solvation. However, if the shape and/or solvation factor alters with particle size, viscosity measurements can be used for determining particle sizes.

If a polymer molecule in solution behaves as a random coil, its average end-to-end distance is proportional to the square root of its extended chain length (see page 25) – i.e. proportional to $M_r^{0.5}$, where M_r is the relative molecular mass. The average solvated volume of the polymer molecule is, therefore, proportional to $M_r^{1.5}$ and, since the unsolvated volume is proportional to M_r, the average solvation factor is proportional to $M_r^{1.5}/M_r$ (i.e. $M_r^{0.5}$). The intrinsic viscosity of a polymer solution is, in turn, proportional to the average solvation factor of the polymer coils – i.e.

$$[\eta] = KM_r^{0.5}$$

where K is a proportionality constant.

For most linear high polymers in solution the chains are somewhat more extended than random, and the relation between intrinsic viscosity and relative molecular mass can be expressed by the general equation proposed by Mark and Houwink:

$$[\eta] = KM_r^{\alpha} \tag{9.6}$$

K and α are constants characteristic of the polymer–solvent system (α depends on the configuration of the polymer chains) and approximately independent of relative molecular mass.

Table 9.1 Values of K and α for some polymer–solvent systems

System	$K/\text{m}^3\text{kg}^{-1}$	α
Cellulose acetate in acetone (25°C)	1.49×10^{-5}	0.82
Polystyrene in toluene (25°C)	3.70×10^{-5}	0.62
Poly(methyl methacrylate) in benzene (25°C	0.94×10^{-5}	0.76
Poly(vinyl chloride) in cyclohexanone (25°C)	0.11×10^{-5}	1.0

In view of experimental simplicity and accuracy, viscosity measurements are extremely useful for routine relative molecular mass determinations on a particular polymer–solvent system. K and α for the system are determined by measuring the intrinsic viscosities of polymer fractions for which the relative molecular masses have been determined independently – e.g. by osmotic pressure, sedimentation or light scattering.

For polydispersed systems an average relative molecular mass intermediate between number-average ($\alpha = 0$) and mass-average ($\alpha = 1$) usually results.

Non-Newtonian flow

Steady-state phenomena

Shear-thinning

Shear-thinning, as the term suggests, is characterised by a gradual (time-independent) decrease in apparent viscosity with increasing rate of shear, and can arise from a number of causes.

If particle aggregation occurs in a colloidal system, then an increase in the shear rate will tend to break down the aggregates, which will result, among other things, in a reduction of the amount of solvent immobilised by the particles, thus lowering the apparent viscosity of the system.

Shear-thinning is particularly common to systems containing asymmetric particles. Asymmetric particles disturb the flow lines to a greater extent when they are randomly orientated at low-velocity gradients than when they have been aligned at high-velocity gradients. In addition, particle interaction and solvent immobilisation are favoured when conditions of random orientation prevail.

The apparent viscosity of a system which thins on shearing is most susceptible to changes in the shear rate in the intermediate range where there is a balance between randomness and alignment, and between aggregation and dispersion.

Plasticity and yield value

Plasticity is similar to shear-thinning, except that the system does not flow noticeably until the shearing stress exceeds a certain minimum value. The applied stress corresponding to a small but arbitrarily chosen rate of deformation is termed the *yield value*.

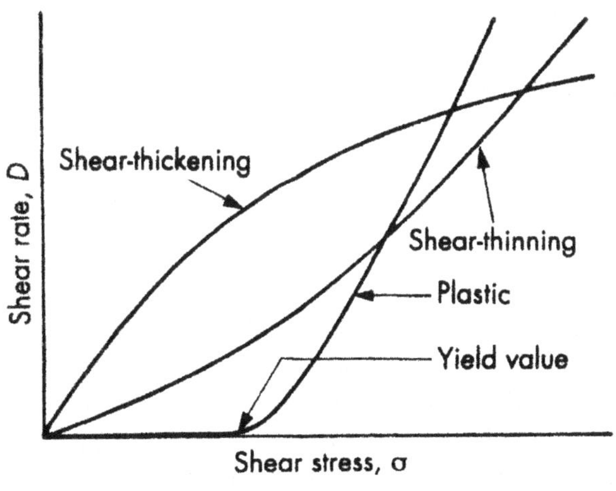

Figure 9.4 Steady-state forms of non-Newtonian flow

Many colloidal systems exhibit so-called Bingham flow, which is characterised by the equation

$$\sigma - \sigma_B = \eta D \tag{9.7}$$

Plasticity is due to a continuous structural network which imparts rigidity to the sample and which must be broken before flow can occur. It is often difficult to distinguish between plastic and ordinary shear-thinning behaviour. Modelling clay, drilling muds and certain pigment dispersions are examples of plastic dispersions. Suspensions of carbon black in hydrocarbon oil often acquire a yield value on standing and become conducting, owing to the contact between the carbon particles which is developed throughout the system[217]. Shearing reduces this conductivity, and the addition of peptising agents reduces both the conductivity and the yield value.

Shear-thickening

Shear-thickening is characterised by an increase in apparent viscosity with increasing rate of deformation.

Shear-thickening is shown in particular, as a dilatant effect, by pastes of densely packed peptised particles in which there is only sufficient liquid to fill the voids. As the shear rate is increased, this dense packing must be broken down to permit the particles to flow past one another. The resulting expansion leaves insufficient liquid to fill the voids and is opposed by surface tension forces. This explains why wet sand apparently becomes dry and firm when walked upon.

Time-dependent phenomena

Thixotropy

Thixotropy is the time-dependent analogue of shear-thinning and plastic behaviour, and arises from somewhat similar causes. If a thixotropic system is allowed to stand and is then sheared at a constant rate, the apparent viscosity decreases with time until a balance between structural breakdown and structure re-formation is reached. If the sheared system is then allowed to stand, it eventually regains its original structure. A thixotropic hysteresis loop (Figure

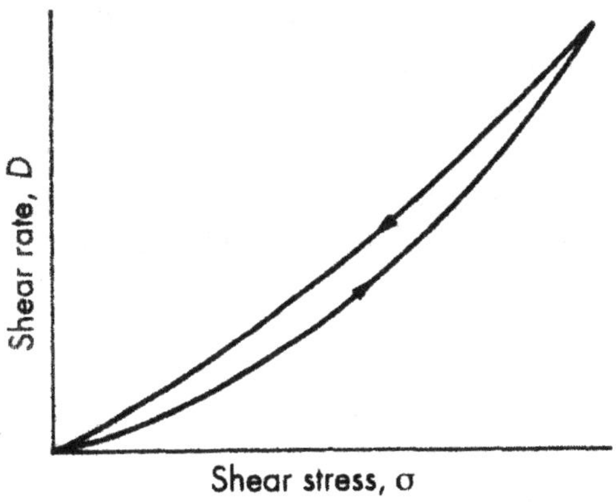

Figure 9.5 A thixotropic loop

9.5) can be obtained by measuring the non-equilibrium shear stress as the shear rate is first increased and then decreased in a standard way.

Solutions of high polymers are, in general, thixotropic to a certain extent; intermolecular attractions and entanglements are overcome and the extent of solvent immobilisation is reduced on shearing, while Brownian motion restores the system to its original condition when left to stand. The classical examples of thixotropic behaviour are given by the weak gel systems, such as flocculated sols of iron(III) oxide, alumina and many clays (particularly bentonite clays), which can be 'liquefied' on shaking and 'solidify' on standing. Thixotropy is particularly important in the paint industry, as it is desirable that the paint should flow only when being brushed on to the appropriate surface (high rate of shear) and immediately after brushing.

Rheopexy

Rheopexy is time-dependent shear-thickening, and is sometimes observed as an acceleration of thixotropic recovery – for example, bentonite clay suspensions often set only slowly on standing but quite rapidly when gently disturbed.

Irreversible phenomena

Shearing sometimes leads to an irreversible breaking (*rheodestruction*) of linkages between the structural elements of a material – e.g. with dehydrated silica gel networks.

Work-hardening can occur as a result of mechanical entanglement or jamming of the structural elements on shearing, an example of this being the 'necking' and corresponding toughening of metal rods when subjected to a tensile stress. A technically important rheological property, which is related to strain-hardening (and to flow elasticity), is *spinability* – i.e. the facility with which a material can be drawn into threads.

Viscoelasticity[122]

When a typical elastic solid is stressed, it immediately deforms by an amount proportional to the applied stress and maintains a constant deformation as long as the stress remains constant – i.e. it obeys Hooke's law. On removal of the stress, the elastic energy stored in the solid is released and the solid immediately recovers its original shape. Newtonian liquids, on the other hand, deform at a rate proportional to the applied stress and show no recovery when the stress is removed, the energy involved having been dissipated as heat in overcoming the internal frictional resistance.

When viscoelastic materials are stressed, some of the energy involved is stored elastically, various parts of the system being deformed into new non-equilibrium positions relative to one another. The remainder is dissipated as heat, various parts of the system flowing into new equilibrium positions relative to one another. If the relative motion of the segments into non-equilibrium positions is hampered, the elastic deformation and recovery of the material is time-dependent (*retarded elasticity*).

Experimental methods

Numerous instruments (plastometers, penetrometers, extensiometers, etc.) and procedures have been devised for measuring the rheological behaviour of various viscoelastic materials. However, the results obtained from most of these instruments are of little fundamental significance, because the applied stress is not uniformly distributed throughout the sample, and the way in which the material behaves towards a particular apparatus is measured rather than a fundamental property of the material itself. Nevertheless, such empirical instruments are indispensable for control testing purposes in industry, where an arbitrary number which bears some relation to the mechanical property under consideration is usually quite sufficient[118].

To measure elastic and viscous properties which are characteristic of the material under consideration and independent of the nature of the apparatus employed, the applied stress and the resulting deformation must be uniform throughout the sample. Concentric cylinder and cone and plate methods approximate these requirements. For materials which are self-supporting, measurements on, for example, the shearing of rectangular samples are ideal.

Creep measurements involve the application of a constant stress (usually a shearing stress) to the sample and the measurement of the resulting sample deformation as a function of time. Figure 9.6 shows a typical creep and recovery curve. In *stress-relaxation* measurements, the sample is subjected to an instantaneous predetermined deformation and the decay of the stress within the sample as the structural segments flow into more relaxed positions is measured as a function of time.

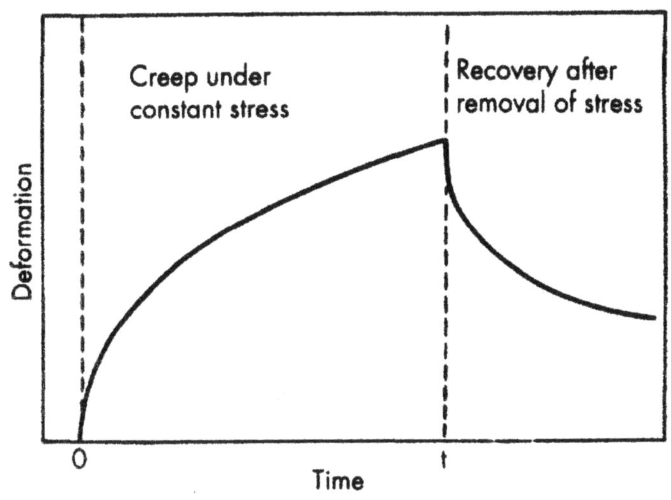

Figure 9.6 Creep and recovery curve for a typical viscoelastic material

The response of a material to an applied stress after very short times can be measured dynamically by applying a sinusoidally varying stress to the sample. A phase difference, which depends on the viscoelastic nature of the material, is set up between stress and strain.

For Hookean elastic solids the stress and strain are in phase, whereas for purely viscous liquids the strain lags 90° behind the applied stress.

Time-dependent deformation and structural characteristics

The deformation of a material when subjected to a constant stress is, as discussed, usually time-dependent. At times of *c*. 10^{-6} s and less all materials, including liquids, have *shear compliances* (i.e shear/shear stress) of *c*. 10^{-11} to 10^{-9} m^2 N^{-1}. This is because there is only sufficient time available for an alteration of interatomic distances and bending of bond angles to take place, and the response of all materials is of the same order of magnitude in this respect. The time required for the various structural units of a material to move into new positions relative to one another depends on the size and shape of the units and the strength of the bonds between them.

The molecules of a liquid start to move relative to one another and the shear compliance increases rapidly after the shearing force has been applied for only *c*. 10^{-6} s. On the other hand, hard solids, such as diamond, sodium chloride crystals and materials at a low enough temperature to be in the glassy state, show only the above rapid elastic deformation, even after the shearing stress has been applied for a considerable time.

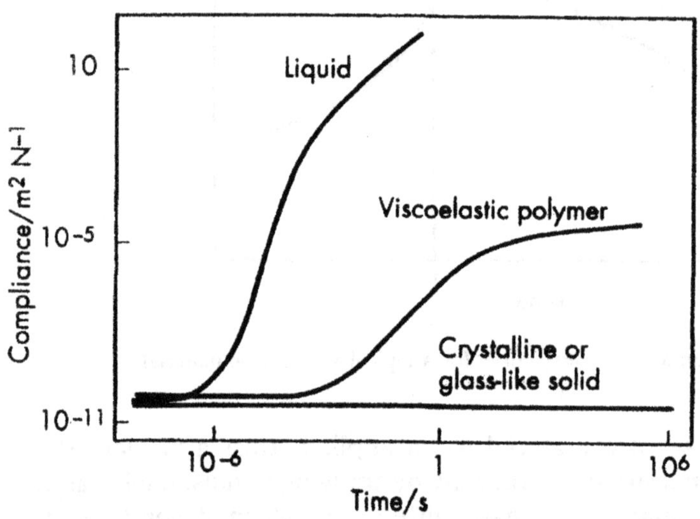

Figure 9.7 The time-dependence of deformations under constant stress

The time taken for the structural units in viscoelastic materials, such as high polymers, to flow into new positions relative to one another is within a few decades of 1 s. Polymer molecules are in a continual state of flexing and twisting, owing to their thermal energy. The configurations of the polymer chains alter more rapidly on a local scale than on a long-range scale. Under the influence of an external stress, the polymer molecules flex and twist into more relaxed positions, again more rapidly on a local scale. In general, there is a continuous range on the time scale covering the response of such systems to external stresses. On this basis, information concerning the structural nature of viscoelastic materials (particularly high polymers) can be obtained by measuring the compliance over a wide range on the time scale[122] (by dynamic methods for times of less than *c*. 1 s, and creep measurements for times of greater than *c*. 1 s).

Rubber elasticity

Rubber-like materials (*elastomers*) have a structure based on polymer chains (e.g. polyisoprene chains, $-CH_2-C(CH_3)=CH-CH_2$ in natural rubber) anchored at various points of cross-linkage. The extent of cross-linking can be increased by vulcanisation. When stretched, the polymer chains are extended lengthwise and compressed crosswise (Figure 9.8) from their average configurations, the change in total volume being inappreciable, and thermal restoring forces are generated. When the tension is released, the polymer chains return (usually rapidly) by thermal motion to their original average configurations.

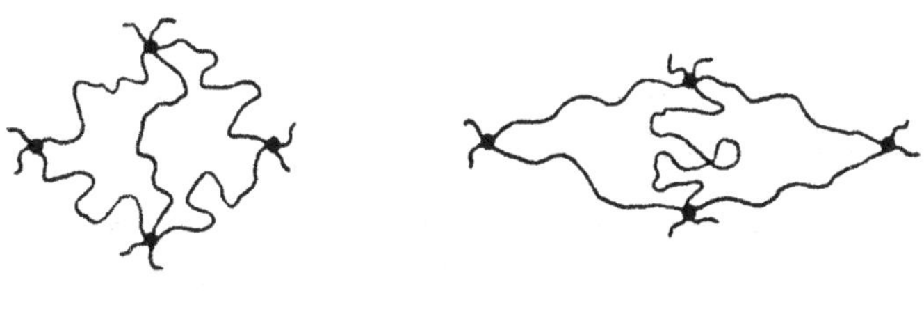

Figure 9.8 The stretching of rubber

Owing to the thermal nature of the restoring force, the deformation of rubber, for a given load, decreases as the temperature is raised. This contrasts with the elasticity of a metal spring, which is due to individual atoms being slightly displaced from their local equilibrium positions, the coil structure greatly multiplying this effect, and which increases with increasing temperature.

If the degree of cross-linking is not very great, as in crude rubber, viscous flow can occur, the polymer chains moving permanently into new equilibrium positions. Excessive cross-linking, on the other hand, restricts changes in the chain configurations and the rubber becomes hard and difficult to deform.

Partial crystallisation may take place in polymeric materials, especially when stretched and/or cooled. From the mechanical standpoint, the introduction of crystalline regions in a polymer is equivalent to increasing the degree of cross-linking, and a partial loss of elasticity results.

Polymers exhibit a glass transition temperature below which the chain arrangements are frozen. Thermal motion no longer overcomes the attractive forces between the polymer chains, and the sample becomes hard and brittle.

Non-linear viscoelasticity

Viscoelasticity is termed *linear* when the time-dependent compliance (strain/stress) of a material is independent of the magnitude of the applied stress. All materials have a linearity limit (see Table 9.2).

Table 9.2 Linear viscoelasticity limits

Material	*Stress*/N m^{-2}	*Percentage strain*
Elastomers	*c*. 10^6–10^7	*c*. 10–100
Plastics	*c*. 10^6–10^7	*c*. 0.1–1
Fats	*c*. 10^2	*c*. 0.01

The linearity limit of elastomers is large, because their deformation is of an entropic nature and does not involve bond rupture and re-formation.

Viscoelastic materials have much lower linearity limits. For the segments or particles in such systems to move (flow) relative to one another without weakening the material, the forces (specific and non-specific) between them must be overcome and then reinstated at the same rate in new positions. If the deforming stress is such that these forces are not reinstated as rapidly as they are overcome, the material becomes structurally weaker. The remaining forces in certain cross-sections between the structural units are then overcome even more readily by the applied stress, and cracks might appear in the sample. Materials with low linear viscoelastic limits are, therefore, those which are readily *work-softened*.

The Weissenberg effect

A characteristic of viscoelastic behaviour is the tendency for flow to occur at right angles to the applied force. An extreme example of this behaviour is illustrated in Figure 9.9. When a rotating rod is lowered into a Newtonian liquid, the liquid is set into rotation and tends to move outwards, leaving a depression around the rod. When the rotating rod is lowered into a viscoelastic liquid, the liquid may actually climb up the rod. The rotation of the rod causes the liquid to be sheared circularly and, because of its elastic nature, it acts like a stretched rubber band, tending to squeeze liquid in towards the centre of the vessel and, therefore, up the rod.

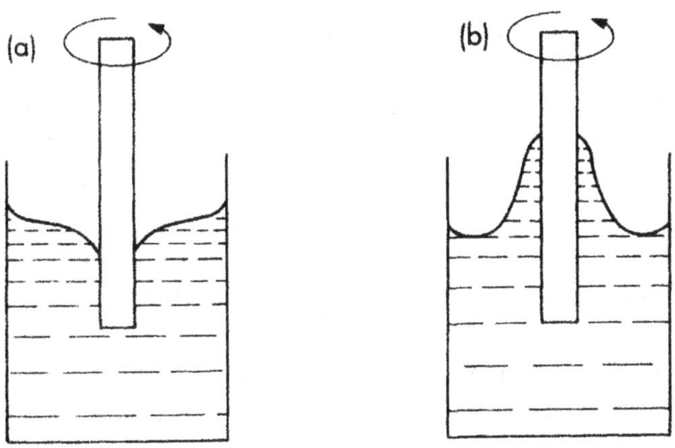

Figure 9.9 The Weissenberg effect: (a) Newtonian liquid; and (b) viscoelastic liquid

10 *Emulsions and foams*

Oil-in-water and water-in-oil emulsions[123–126]

An emulsion is a dispersed system in which the phases are immiscible or partially miscible liquids. The globules of the dispersed liquid in the usual type of emulsion (sometimes now called a macroemulsion) are usually between 0.1 μm and 10 μm in diameter, and so tend to be larger than the particles found in sols.

The practical application of emulsions and emulsion technology is considerable, and includes foodstuffs (especially dairy produce), pharmaceutical preparations, cosmetics, agricultural sprays and bituminous products. Emulsions enable the dilution of an expensive or concentrated ingredient with an inexpensive, but immiscible, diluent. For example, water-insoluble agrochemicals are generally marketed in the form of oil in water emulsions which can be diluted with water by the user in order to permit even, low-level application. Emulsions also allow enhanced control over other factors, such as rheology and the various manifestations of chemical reactivity (including taste, odour and toxicity). A large volume of technological information on emulsions exists, much of it in private files. In general, empirical knowledge is well ahead of fundamental understanding.

Microemulsions, with droplet diameters of 0.01–0.1 μm, can also be prepared and these are currently the subject of much fundamental investigation and new applications (e.g. oil recovery from porous rocks). Whether microemulsions should be regarded as true emulsions or as swollen micelles (see section on solubilisation, page 89) is a matter of controversy.

The visual appearance of an emulsion reflects the influence of droplet size on light scattering, and varies from milky-white-opaque, with large droplets, through blue-white, then gray-translucent, to transparent, with small microemulsion droplets.

In nearly all emulsions, one of the phases is aqueous and the other is (in the widest sense of the term) an oil. If the oil is the dispersed phase, the emulsion is termed an *oil-in-water* (O/W) emulsion; if the aqueous medium is the dispersed phase, the emulsion is termed a *water-in-oil* (W/O) emulsion. There are several methods by which the emulsion type may be identified.

1. In general, an O/W emulsion has a creamy texture and a W/O emulsion feels greasy.
2. The emulsion mixes readily with a liquid which is miscible with its dispersion medium.
3. The emulsion is readily coloured by dyes which are soluble in the dispersion medium.
4. O/W emulsions usually have a much higher electrical conductivity than W/O emulsions.

Emulsifying agents and emulsion stability

Probably the most important physical property of an emulsion is its stability. The term 'emulsion stability' can be used with reference to three essentially different phenomena – creaming (or sedimentation), coagulation and a breaking of the emulsion due to droplet coalescence.

Creaming results from a density difference between the two phases and is not necessarily accompanied by droplet coagulation, although it facilitates this process.

Droplet collisions may result in coagulation, which, in turn, may lead to coalescence into larger globules. Eventually, the dispersed phase may become a continuous phase, separated from the dispersion medium by a single interface. The time taken for such phase separation may be anything from seconds to years, depending on the emulsion formulation and manufacturing conditions.

Assessment of the stability of an emulsion against coalescence involves droplet counting[218]. The most unequivocal method (but one which is rather laborious) is to introduce a suitably diluted sample of the emulsion into a haemocytometer cell and count the microscopically visible particles manually.

The Coulter counter affords a convenient indirect technique for determining droplet size distributions in O/W emulsions. In this method a dilute dispersion is made to flow through a small orifice. The passage of a non-conducting particle through the orifice causes a momentary increase in the electric resistance between electrodes placed either side of the orifice. The magnitude of this increase in resistance depends on the size of the particle. The electronic circuitry of the system is such that particles above a certain size are counted every time the resistance rises above the corresponding preselected cut-off value. By varying this cut-off value, a cumulative particle size distribution can be determined. The lower particle diameter limit for this technique is about 0.7 μm, which makes it unsuitable for the study of most colloidal dispersions, but suitable for studying the relatively large droplets in O/W emulsions.

If an emulsion is prepared by homogenising two pure liquid components, phase separation will usually be rapid, especially if the concentration of the dispersed phase is at all high. To prepare reasonably stable emulsions, a third component – an *emulsifying agent* (or *emulsifier*) – must be present. The materials which are most effective as emulsifying (and foaming) agents can be broadly classified as:

1. Surface-active materials.
2. Naturally occurring materials.
3. Finely divided solids.

The functions of the emulsifying agent are to facilitate emulsification and promote emulsion stability. The emulsifying agent forms an adsorbed film around the dispersed droplets which helps to prevent coagulation and coalescence. The stabilising mechanism is usually complex and may vary from system to system. In general, however, the factors which control droplet coagulation are the same as those which control the stability of sols (see Chapter 8), whereas stability against droplet coalescence depends mainly on the mechanical properties of the interfacial film.

The following factors (which depend on the nature of the emulsifying agent and/or on a suitable choice of formulation and manufacturing conditions) favour emulsion stability:

1. *Low interfacial tension* The adsorption of surfactant at oil–water interfaces causes a lowering of interfacial energy, thus facilitating

the development and enhancing the stability of the large interfacial areas associated with emulsions.

2. *A mechanically strong and elastic interfacial film* This is particularly important when the volume fraction of the dispersed phase is high.

   The stability of emulsions stabilised by proteins arises from the mechanical protection given by the adsorbed films around the droplets rather than from a reduction of interfacial tension.

   Finely divided solids for which the contact angle is between 0° and 180° have a tendency to collect at the oil–water interface (cf. flotation; page 161), where they impart stability to the emulsion.

   Surfactants can also stabilise in the mechanical sense. Coalescence involves droplet coagulation followed by a squeezing of film material from the region of droplet contact, and the latter is more favoured with an expanded film than with a close-packed film. For example, very stable hydrocarbon oil in water emulsions can be prepared with sodium cetyl sulphate (dissolved in the water) plus cetyl alcohol (dissolved in the oil) as emulsifier (a condensed mixed film being formed at the interface), whereas hydrocarbon oil in water emulsions prepared with sodium cetyl sulphate plus oleyl alcohol (which gives an expanded mixed film) are much less stable[219]. The most effective interfacial films are the mixed films which are formed as a result of the combined use of water-soluble and oil-soluble emulsifying agents.

   It is also important that the emulsifier films have sufficient elasticity to enable recovery from local disturbances (see Gibbs–Marangoni effect; page 274).

3. *Electrical double layer repulsions* (see page 212) Interparticle repulsion due to the overlap of similarly charged electric double layers is an important stabilising mechanism in O/W emulsions.

   When ionic emulsifying agents are used, lateral electric double layer repulsion may prevent the formation of a close-packed film. This film-expanding effect can be minimised by using a mixed ionic plus non-ionic film[220] (see above) and/or by increasing the electrolyte concentration in the aqueous phase[221].

4. *Relatively small volume of dispersed phase* (see below).

5. *Narrow droplet size distribution* Larger droplets are less unstable than smaller droplets on account of their smaller area-to-volume ratio, and so will tend to grow at the expense of the smaller droplets (see page 68). If this process continues, the emulsion will

eventually break. Emulsions with a fairly uniform droplet size will be less prone to this effect.

6. *High viscosity* A high Newtonian viscosity simply retards the rates of creaming, coalescence, etc. If a weak gel network is formed by, for example, dissolving sodium carboxymethyl cellulose in an O/W emulsion, genuine stability might ensue. However, the overall rheological properties of such an emulsion may not be acceptable.

Emulsifying agents and emulsion type

The type of emulsion which is formed when a given pair of immiscible liquids is homogenised depends on (1) the relative volumes of the two phases, and (2) the nature of the emulsifying agent.

1. *Phase volume* The higher its phase volume, the more likely a liquid is to become the dispersion medium. However, the liquid with the greater phase volume need not necessarily be the dispersion medium.

   If the emulsion consisted of an assembly of closely packed uniform spherical droplets, the dispersed phase would occupy 0.74 of the total volume. Stable emulsions can, however, be prepared in which the volume fraction of the dispersed phase exceeds 0.74, because (*a*) the droplets are not of uniform size and can, therefore, be packed more densely, and (*b*) the droplets may be deformed into polyhedra, the interfacial film preventing coalescence.
2. *Nature of the emulsifying agent* Alkali-metal soaps favour the formation of O/W emulsions, whereas heavy-metal soaps favour the formation of W/O emulsions. O/W emulsions in the middle concentration region stabilised by alkali-metal soaps can often be broken, and even inverted into W/O emulsions, by the addition of heavy-metal ions.

Several theories relating to emulsion type have been proposed. The most satisfactory general theory of emulsion type is that originally proposed for emulsions stabilised by finely divided solids (see Figure 10.1). If the solid is preferentially wetted by one of the phases, then more particles can be accommodated at the interface if the interface is convex towards that phase (i.e. if the preferentially

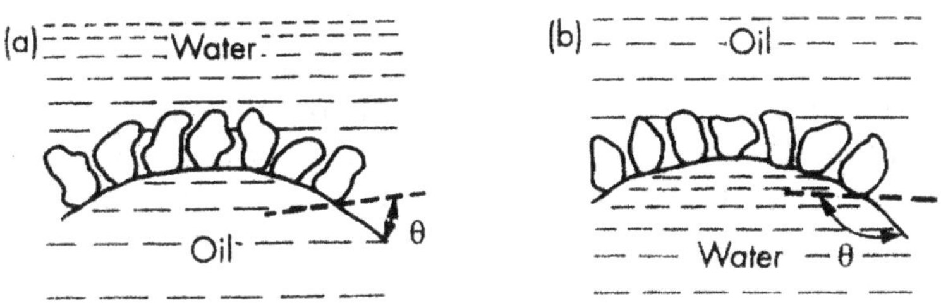

Figure 10.1 Stabilisation of emulsions by finely divided solids: (a) preferential wetting by water leading to an O/W emulsion; (b) preferential wetting by oil leading to a W/O emulsion

wetting phase is the dispersion medium). For example, bentonite clays (which are preferentially wetted by water) tend to give O/W emulsions, whereas carbon black (which is preferentially oil-wetted) tends to give W/O emulsions. This preferential wetting theory can be extended to cover other types of emulsifying agent. The type of emulsion which tends to form depends on the balance between the hydrophilic and the lipophilic properties of the emulsifier – alkali-metal soaps favour the formation of O/W emulsions because they are more hydrophilic than lipophilic, whereas the reverse holds for heavy-metal soaps.

The generalisation that 'the phase in which the emulsifying agent is the more soluble tends to be the dispersion medium' is known as Bancroft's rule.

The amphiphilic nature of many emulsifying agents (particularly non-ionic surfactants) can be expressed in terms of an empirical scale of so-called HLB (hydrophile–lipophile balance) numbers[222] (see Table 10.1). The least hydrophilic surfactants are assigned the lowest HLB values. Several formulae have been established for calculating HLB numbers from composition data and they can also be determined experimentally – e.g. from cloud-point measurements[123,125]. For mixed emulsifiers, approximate algebraic additivity holds.

The optimum HLB number for forming an emulsion depends to some extent on the nature of the particular system. Suppose that 20 per cent sorbitan tristearate (HLB 2.1) plus 80 per cent polyoxyethylene sorbitan monostearate (HLB 14.9) is the optimum composition of a mixture of these emulsifiers for preparing a particular O/W emulsion. The HLB of the mixture is, therefore, $(0.2 \times 2.1) + (0.8 \times 14.9) = 12.3$. The theory is that an HLB of 12.3 should be optimum for the formation of this particular O/W emulsion by use of other emulsifier systems; for example, the optimum proportions in a mixture of sorbitan mono-oleate (HLB 4.3) and polyoxyethylene sorbitan monopalmitate (HLB 15.6) should be approximately 30 per cent and 70 per cent, respectively. In commercial emulsion formulation, HLB numbers are used advantageously in this way as an initial guide prior to a certain amount of trial and error testing.

Table 10.1 HLB values

Applications		*Dispersibility in water*	
3–6	W/O emulsions	1–4	Nil
7–9	Wetting agents	3–6	Poor
8–15	O/W emulsions	6–8	Unstable milky dispersion
13–15	Detergent	8–10	Stable milky dispersion
15–18	Solubiliser	10–13	Translucent dispersion/solution
		13–	Clear solution

A major disadvantage of the HLB concept is that it makes no allowance for temperature effects. With increasing temperature, the hydration of lyophilic (particularly poly(ethylene oxide)) groups decreases and the emulsifying agent becomes less hydrophilic – i.e. its HLB decreases.

An alternative method for characterising emulsifying agents is in terms of a phase inversion temperature (PIT)[223]. An emulsion containing equal weights of oil and water phases and 3–5 per cent emulsifier is heated and agitated, and the temperature at which the emulsion inverts is noted; i.e. the emulsion is W/O above and O/W below the PIT. Therefore, the PIT should be higher than the proposed storage temperature for an O/W emulsion and lower than the proposed storage temperature for a W/O emulsion.

O/W emulsions with a very small droplet size can be obtained if prepared at only a few kelvins below the PIT. At this temperature the emulsion will be unstable to coalescence, but subsequent cooling to 20 K or more below the PIT can enhance the stability of the emulsion while retaining the small average droplet size[224].

Breaking of emulsions

In many instances it is the breaking of an emulsion (demulsification) which is of practical importance. Examples are the creaming, breaking and inversion of milk to obtain butter, and the breaking of W/O oil-field emulsions. Small amounts of water often get emulsified in lubricating oils, hydraulic oils and heat-exchange systems, and it is necessary to remove this water to prevent corrosion and other undesirable effects.

A number of techniques are used commercially to accelerate emulsion breakdown. Mechanical methods include centrifugal separation, freezing, distillation and filtration. Another method is based on the principle of antagonistic action – i.e. the addition of O/W-promoting emulsifiers tends to break W/O emulsions, and vice-versa. Emulsions can also be broken by the application of intense electrical fields, the principal factors involved being electrophoresis in the case of O/W emulsions and droplet deformation in the case of W/O emulsions.

Microemulsions[127–128]

Microemulsions are emulsions with droplet diameters in the range of about 0.01 to 0.1 μm. They are, consequently, of low turbidity.

The formation of a microemulsion involves the creation of a situation in which the oil–water interfacial tension approaches zero (or even becomes transiently negative). With all single ionic surfactants and most single non-ionic surfactants this is not possible, since γ_{OW} is still sizeable when the c.m.c. or the limit of solubility is reached. To achieve the required lowering of γ_{OW} a co-surfactant must be included. For example, O/W microemulsions can be formed using a mixture of potassium oleate and pentanol as the emulsifying agent. In general, less co-surfactant is required for O/W than for W/O microemulsions. Electrolytes promote the formation of W/O microemulsions.

In view of the high oil–water interfacial area which must be created, the fraction of emulsifying agent in microemulsion formulations tends to be significantly higher than that in ordinary emulsions. A typical microemulsion formulation would be 10–70 per cent oil, 10–70 per cent water and 5–40 per cent emulsifying agent[48].

With γ_{OW} close to zero, microemulsions will form spontaneously and are thermodynamically stable. The droplets of microemulsions tend to be monodispersed. A microemulsion may form as a separate phase in equilibrium with excess oil (O/W) or water (W/O) (i.e. it is saturated with respect to droplets). Microemulsions are usually of low viscosity.

Microemulsions represent an intermediate state between micelles and ordinary emulsions, and it is a debatable issue whether or not they should be considered as swollen micelles rather than as small-droplet emulsions. Droplet size, though small, is nonetheless large enough to justify classification as emulsions. On the other hand, the observed thermodynamic stability and reproducibility is uncharacteristic of ordinary emulsions.

Microemulsions are potentially exploitable in any situation where the mixing of oil and water is desired. The possibility of using them to enhance tertiary oil recovery has recently attracted a great deal of attention.

The recovery of oil from natural reservoirs involves three stages. In the primary stage, oil is forced out of the reservoir by the pressure of natural gases. When this pressure is no longer adequate, the secondary stage is effected in which water is pumped into the reservoir to force out further oil. This still usually leaves well over half of the total oil unrecovered, most of it being trapped in the pore structure of the reservoir by capillary and viscous forces. In the tertiary stage, some of this residual oil can be recovered by injection of oil-miscible fluids and/or forced out by steam under high pressure. The injection of appropriate microemulsifying surfactant mixtures offers the possibility of significantly enhancing this final stage of oil recovery.

Foams[129–130]

A foam is a coarse dispersion of gas in liquid, and two extreme structural situations can be recognised. The first type (dilute foams) consist of nearly spherical bubbles separated by rather thick films of somewhat viscous liquid. The other type (concentrated foams) are mostly gas phase, and consist of polyhedral gas cells separated by thin liquid films (which may develop from more dilute foams as a result of liquid drainage, or directly from a liquid of relatively low viscosity)*. The nature of thin liquid films (as found in these concentrated foams) has been the subject of a great deal of fundamental research.

Foam stability

Only transitory foams can be formed with pure liquids and, as with emulsions, a third (surface-active) component – a *foaming agent* – is necessary to achieve any reasonable degree of stability. Good emulsifying agents are, in general, also good foaming agents, since the factors which influence emulsion stability (against droplet coalescence) and foam stability (against bubble collapse) are somewhat similar.

The stability of a foam depends upon two principal factors – the tendency for the liquid films to drain and become thinner, and their tendency to rupture as a result of random disturbances. Other factors which may significantly influence foam stability include evaporation and gas diffusion through the liquid films. Owing to their high interfacial area (and surface free energy), all foams are unstable in the thermodynamic sense. Some distinction can be made, however, between *unstable* and *metastable* foam structures. *Unstable* foams are typified by those formed from aqueous solutions of short-chain fatty acids or alcohols. The presence of these mildly surface-active agents retards drainage and film rupture to some extent, but does not stop these processes from continuously taking place to the point of complete foam collapse. *Metastable* foams are typified by those formed from solutions of soaps, synthetic detergents, proteins, saponins, etc. The balance of forces is such that the drainage of liquid stops when a certain film thickness is reached and, in the absence of disturbing influences (such as vibration, draughts, evaporation, diffusion of gas from small bubbles to large bubbles, heat, temperature gradients, dust and other impurities), these foams would persist almost indefinitely.

*Similarly, some solid foams (e.g. foam rubber) consist of spherical gas bubbles trapped within a solid network, whereas others (e.g. expanded polystyrene) consist of as little as 1 per cent solid volume and are composed of polyhedral gas cells separated by very thin solid walls.

Foam drainage

Initially, the liquid films of a foam will be relatively thick and drainage will take place mainly by gravitational flow throughout the whole of these films. The foaming agent plays an important role even at this stage in restricting this flow to a level where local disturbances and consequent film rupture is minimised. When the films have attained a thickness of the order of micrometres, gravitational flow down the laminar regions will become extremely slow. The predominant drainage mechanism will then involve liquid being discharged locally at positions of inter-film contact (known as Plateau borders), where the liquid capacity is relatively high. Subsequent drainage then takes place through the network of Plateau borders that exists throughout the foam.

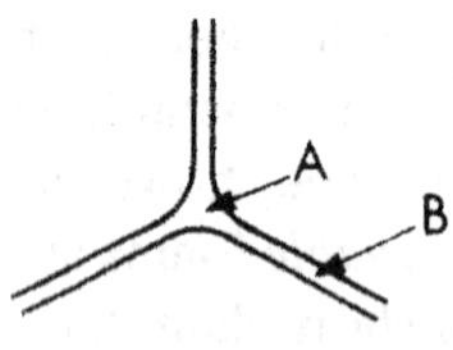

Figure 10.2 Plateau border at a line of intersection of three bubbles. Owing to the curvature of the liquid–gas interface at A, the pressure of liquid at A is lower than at B, thus causing capillary flow of liquid towards A

The discharge of liquid from the laminar part of a thin film is governed by the pressure of the liquid in this region compared with that of the liquid in the Plateau borders. At least three factors are likely to be involved. van der Waals attractive forces favour film thinning, and the overlapping of similarly charged electric double layers opposes film thinning (see Chapter 8). The other important factor is a capillary pressure, which favours film thinning. This arises because the pressure of the adjacent gas phase is uniform and, therefore, the pressure of liquid in the Plateau borders, where the interface is curved, is less than that of liquid in the laminar film. Depending on the balance of these forces, a film may either thin continuously and eventually rupture, or attain an equilibrium thickness. Any propagated structure within the film may significantly affect the equilibrium film thickness that a balance between the above forces would otherwise determine.

Experimental studies on non-draining, horizontal liquid films in which the equilibrium film thickness is measured as a function of ionic strength and applied hydrostatic pressure (or suction) provides a means of investigating the above forces[98,225–226]. Figure 10.3 shows an apparatus used by Deryagin and Titijevskaya[225]. A flat liquid film with an area of *c.* 1 mm^2 is formed between cups A and B which are connected via tube C to equalise bubble pressures. The pressure in the bubbles in excess of that in the liquid film is calculated from the manometer reading Δh. Deryagin called this excess pressure the *disjoining pressure*. An elaborate optical device (not shown) permits measurement of the film thickness.

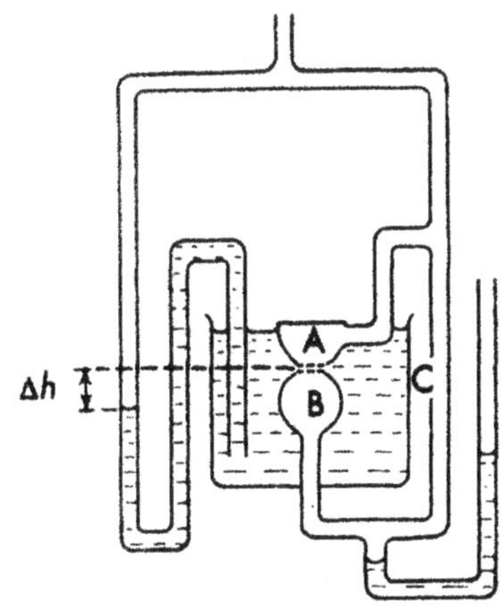

Figure 10.3 Apparatus for measuring the disjoining pressure of free films as a function of their thickness[225] (After B.V. Deryagin and A.S. Titijevskaya)

Figure 10.4 shows the results of some measurements on aqueous sodium oleate films. The sensitivity of the equilibrium film thickness to added electrolyte reflects qualitatively the expected positive contribution of electric double layer repulsion to the disjoining pressure. However, this sensitivity to added electrolyte is much less than that predicted from electric double layer theory and at high electrolyte concentration an equilibrium film thickness of *c.* 12 nm is attained which is almost independent of the magnitude of the disjoining pressure. To account for this observation, Deryagin and Titijevskaya have postulated the existence of hydration layers

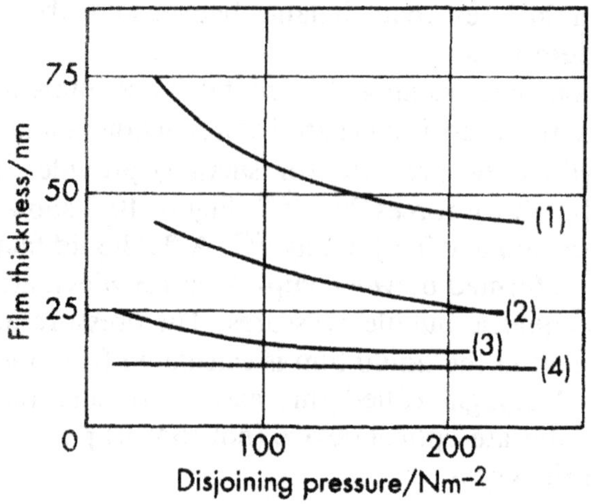

Figure 10.4 Film thickness as a function of disjoining pressure for films of 10^{-3} mol dm^{-3} aqueous sodium oleate containing NaCl at concentrations of (1) 10^{-4} mol dm^{-3}; (2) 10^{-3} mol dm^{-3}; (3) 10^{-2} mol dm^{-3}; and (4) 10^{-1} mol dm^{-3} (After B.V. Deryagin and A.S. Titijevskaya[225])

effectively *c.* 6 nm thick. Similar conclusions have also been reached by van den Tempel[227] from experiments with oil droplets in aqueous sodium dodecyl sulphate plus sodium chloride.

Film rupture

In addition to film drainage, the stability of a foam depends on the ability of the liquid films to resist excessive local thinning and rupture which may occur as a result of various random disturbances. A number of factors may be involved with varying degrees of importance, depending on the nature of the particular foam in question.

Gibbs–Marangoni surface elasticity effect

This is an important stabilising effect in foams which are formed from solutions of soaps, detergents, etc. If a film is subjected to local stretching as a result of some external disturbance, the consequent increase in surface area will be accompanied by a decrease in the surface excess concentration of foaming agent and, therefore, a local increase in surface tension (Gibbs effect). Since a certain time is required for surfactant molecules to diffuse to this surface region and restore the original surface tension (Marangoni effect), this increased surface tension may persist for long enough to cause the disturbed film region to recover its original thickness[228].

As an extension of the Marangoni effect, Ewers and Sutherland[229] have suggested a surface transport mechanism in which the surface tension gradient created by local film thinning causes foaming agent to spread along the surface and drag with it a significant amount of underlying solution, thus opposing the thinning process.

An absence of the Gibbs–Marangoni effect is the main reason why pure liquids do not foam. It is also interesting, in this respect, to observe that foams from moderately concentrated solutions of soaps, detergents, etc., tend to be less stable than those formed from more dilute solutions. With the more concentrated solutions, the increase in surface tension which results from local thinning is more rapidly nullified by diffusion of surfactant from the bulk solution. The opposition to fluctuations in film thickness by corresponding fluctuations in surface tension is, therefore, less effective.

Surface rheology

The mechanical properties of the surface films (as in the case of emulsions) often have a considerable influence on foam stability. Several considerations may be involved.

A high bulk liquid viscosity simply retards the rate of foam collapse. High surface viscosity, however, involves strong retardation of bulk liquid flow close to the surfaces and, consequently, the drainage of thick films is considerably more rapid than that of thin films, which facilitates the attainment of a uniform film thickness.

Surface elasticity facilitates the maintenance of a uniform film thickness, as discussed above; however, the existence of rigid, condensed surface films is detrimental to foam stability, owing to the very small changes in area over which such films show elasticity.

Equilibrium film thickness

If the balance of van der Waals attraction, electric double layer repulsion, capillary pressure, structure propagation, etc., favours an equilibrium film thickness, random fluctuations in film thickness will, in any case, tend to be neutralised.

Antifoaming agents

The prevention of foaming and the destruction of existing foams is often a matter of practical importance; for example, polyamides and silicones find use as foam inhibitors in water boilers. Antifoaming agents act against the various factors which promote foam stability (described above) and, therefore, a number of mechanisms may be operative.

Foam inhibitors are, in general, materials which tend to be adsorbed in preference to the foaming agent, yet do not have the requisites to form a stable foam. They may be effective by virtue of rapid adsorption; for example, the addition of tributyl phosphate to aqueous sodium oleate solutions significantly reduces the time required to reach equilibrium surface tension[230], thus lessening the Marangoni surface elasticity effect and the foam stability. They may also act, for example, by reducing electric double layer repulsion or by facilitating drainage by reducing hydrogen bonding between the surface films and the underlying solution.

Foams can often be broken by spraying with small quantities of substances such as ether and *n*-octanol. As a result of their high surface activity, these foam breakers raise the surface pressure over small regions of the liquid films and spread from these regions, displacing the foaming agent and carrying with them some of the underlying liquid[229]. Small regions of film are, therefore, thinned and left without the properties to resist rupture.

Problems

$$k = 1.380\,5 \times 10^{-23}\ \text{J K}^{-1}$$
$$N_A = 6.022\,5 \times 10^{23}\ \text{mol}^{-1}$$
$$R = 8.314\,3\ \text{J K}^{-1}\ \text{mol}^{-1}$$
$$e = 1.602\,1 \times 10^{-19}\ \text{C}$$
$$\epsilon_0 = 8.854\,2 \times 10^{-12}\ \text{F m}^{-1}$$
$$g = 9.806\,6\ \text{m s}^{-2}$$

Volume of ideal gas at s.t.p. (0°C and 1 atm) $= 2.241\,4 \times 10^{-2}\ \text{m}^3\ \text{mol}^{-1}$

$$1\ \text{atm} = 760\ \text{mmHg} = 1.013\,25 \times 10^5\ \text{Pa}$$
$$0°\text{C} = 273.15\ \text{K}$$
$$\ln 10 = 2.302\,6$$
$$\pi = 3.141\,56$$
$$\eta\ (\text{water, 25°C}) = 8.9 \times 10^{-4}\ \text{kg m}^{-1}\ \text{s}^{-1}$$
$$\epsilon/\epsilon_0\ (\text{water, 25°C}) = 78.5$$

1. Calculate the average displacement in 1 min along a given axis produced by Brownian motion for a spherical particle of radius 0.1 μm suspended in water at 25°C.

2. The sedimentation and diffusion coefficients for myoglobin in dilute aqueous solution at 20°C are 2.04×10^{-13} s and 1.13×10^{-10} $\text{m}^2\ \text{s}^{-1}$, respectively. The partial specific volume of the protein is 0.741 $\text{cm}^3\ \text{g}^{-1}$, the density of the solution is 1.00 g cm^{-3} and the coefficient of viscosity of the solution is 1.00×10^{-3} kg $\text{m}^{-1}\ \text{s}^{-1}$. Calculate (*a*) the relative molecular mass and (*b*) the frictional ratio of this protein. What is the probable shape of a dissolved myoglobin molecule?

3. An aqueous solution of β-lactoglobulin in the presence of sufficient electrolyte to eliminate charge effects was centrifuged to

equilibrium at 11 000 revolutions per minute and at 25°C. The following equilibrium concentrations were measured:

Distance from axis of rotation/cm	4.90	4.95	5.00	5.05	5.10	5.15
Concentration/g dm^{-3}	1.30	1.46	1.64	1.84	2.06	2.31

The partial specific volume of the protein was 0.75 cm^3 g^{-1} and the density of the solution (assumed constant) was 1.00 g cm^{-3}. Calculate the relative molecular mass of the protein.

4. The following osmotic pressures were measured for solutions of a sample of polyisobutylene in benzene at 25°C:

Concentration/(g/100 cm^3)	0.5	1.0	1.5	2.0
Osmotic pressure/(cm *solution*)	1.03	2.10	3.22	4.39

(solution density = 0.88 g cm^{-3} in each case)

Calculate an average relative molecular mass.

5. A 0.01 mol dm^{-3} aqueous solution of a colloidal electrolyte (which can be represented as $Na_{15}X$) is on one side of a dialysis membrane, while on the other side of the membrane is an equal volume of 0.05 mol dm^{-3} aqueous sodium chloride. When Donnan equilibrium is established, what net fraction of the NaCl will have diffused into the compartment containing the colloidal electrolyte?

6. The following light-scattering data give values of the quantity $10^7\ Kc/R_\theta$ for solutions of cellulose nitrate dissolved in acetone:

Concentration/ g dm^{-3}	*Scattering angle (in relation to transmitted beam)*		
	45°	32°	17° 30′
0.88	69.8	49.0	33.0
0.64	66.0	45.5	29.4
0.43	62.8	42.1	25.9

Make appropriate extrapolations and calculate an average relative molecular mass for the cellulose nitrate.

7. If bubbles of air 10^{-7} m in diameter and no other nuclei are present in water just below the boiling point, by approximately how much could water be superheated at normal atmospheric pressure before boiling starts? The surface tension of water at 100°C is 59 mN m^{-1} and its enthalpy of vaporisation is 2.25 kJ g^{-1}.

8. The following surface tensions were measured for aqueous solutions of the non-ionic surfactant $CH_3(CH_2)_9(OCH_2CH_2)_5OH$ at 25°C:

$c/10^{-4}$ mol dm^{-3}	0.1	0.3	1.0	2.0	5.0	8.0	10.0	20.0	30.0
γ/mN m^{-1}	63.9	56.2	47.2	41.6	34.0	30.3	29.8	29.6	29.5

Determine the critical micelle concentration and calculate the area occupied by each adsorbed surfactant molecule at the critical micelle concentration.

9. The following surface tensions were measured for aqueous solutions of *n*-pentanol at 20°C:

c/mol dm^{-3}	0	0.01	0.02	0.03	0.04	0.05	0.06	0.08	0.10
γ/mN m^{-1}	72.6	64.6	60.0	56.8	54.3	51.9	49.8	46.0	43.0

Calculate surface excess concentrations and the average area occupied by each adsorbed molecule for bulk concentrations of 0.01, 0.02, 0.04 and 0.08 mol dm^{-3}. Plot a π–A curve for the adsorbed *n*-pentanol monolayer and compare it with the corresponding curve for an ideal gaseous film.

10. The surface tensions of aqueous solutions of sodium dodecyl sulphate at 20°C are as follows:

c/mmol dm^{-3}	0	2	4	5	6	7	8	9	10	12
γ/mN m^{-1}	72.0	62.3	52.4	48.5	45.2	42.0	40.0	39.8	39.6	39.5

Calculate the area occupied by each adsorbed dodecyl sulphate ion when adsorption is a maximum and explain the discontinuity in the surface tension–concentration relationship.

11. The c.m.c. of a non-ionic surfactant in water varies with temperature as follows:

Temperature/°C	15	20	25	30	40
c.m.c./mmol dm^{-3}	21.9	20.8	19.9	19.1	17.8

Determine an enthalpy of micellisation in accord with the phase separation model.

12. At 20°C the surface tensions of water and *n*-octane are 72.8 and 21.8 mN m^{-1}, respectively, and the interfacial tension of the *n*-octane–water interface is 50.8 mN m^{-1}. Calculate

(*a*) the work of adhesion between *n*-octane and water;
(*b*) the work of cohesion for (*i*) *n*-octane and (*ii*) water;
(*c*) the initial spreading coefficient of *n*-octane on water;
(*d*) the dispersion component of the surface tension of water.

13. The following surface pressure measurements were obtained for a film of haemoglobin spread on 0.01 mol dm^{-3} HCl (aq.) at 25°C:

A/m^2 mg^{-1}	4.0	5.0	6.0	7.5	10.0
π/mN m^{-1}	0.28	0.16	0.105	0.06	0.035

Calculate the relative molecular mass of the spread protein and compare your answer with a value of 68 000 as determined from sedimentation measurements.

14. The amount of hydrogen chemisorbed on to a nickel surface at 600 K is approximately proportional to the square root of its pressure. Derive a two-dimensional equation of state for the adsorbed gas and comment on its significance.

15. The following data list volumes of ammonia (reduced to s.t.p.) adsorbed by a sample of activated charcoal at 0°C:

Pressure/kPa	6.8	13.5	26.7	53.1	79.4
Volume/cm^3 g^{-1}	74	111	147	177	189

Show that the data fit a Langmuir adsorption isotherm expression and evaluate the constants.

16. The following data refer to the adsorption of nitrogen on 0.92 g of a sample of silica gel at 77 K, p being the pressure and V the volume adsorbed:

p/kPa	3.7	8.5	15.2	23.6	31.5	38.2	46.1	54.8
V/cm^3 (s.t.p.)	82	106	124	142	157	173	196	227

Saturated vapour pressure (p_0) = 101.3 kPa.

Plot the adsorption isotherm and use the BET equation to calculate a specific surface area for the silica gel sample, taking the molecular area of nitrogen as 16.2×10^{-20} m^2.

17. The following data refer to the adsorption of *n*-butane at 273 K by a sample of tungsten powder which has a specific surface area (as determined from nitrogen adsorption measurements at 77 K) of 6.5 m^2 g^{-1}:

Relative pressure (p/p_0)	0.04	0.10	0.16	0.25	0.30	0.37
Volume of gas adsorbed/ cm^3 (s.t.p.) g^{-1}	0.33	0.46	0.54	0.64	0.70	0.77

Use the BET equation to calculate a molecular area for the adsorbed butane at monolayer coverage and compare it with the value of 32.1×10^{-20} m^2 estimated from the density of liquid butane.

18. At 90 K a 1.21 g sample of a porous solid showed the following results for the adsorption of krypton:

Pressure/mmHg	1.110	3.078
Volume of Kr *adsorbed*/cm^3 (s.t.p.)	1.475	1.878

If the saturation vapour pressure and molecular area of krypton at this temperature are 19.0 mmHg and 21×10^{-20} m^2, respectively, calculate a specific surface area for the solid.

19. The following results refer to the adsorption of nitrogen on a graphitised sample of carbon black, and give the ratio of the nitrogen pressures for temperatures of 90 K and 77 K which are required to achieve a given amount of adsorption:

Amount of N_2 *adsorbed* (V/V_m)	0.4	0.8	1.2
p (90 K)/p (77 K)	14.3	17.4	7.8

Calculate an isosteric enthalpy of adsorption for each value of V/V_m and comment on the values obtained.

20. In adsorption studies of a particular system, it was found that the adsorption isotherms could be expressed by empirical equations of the form, $\theta = A\ (p/\text{mmHg})^n$, where θ is the fraction of the solid surface covered by adsorbed gas at equilibrium pressure, p. Studies at two temperatures yielded the following values for the constants of the above equation:

T/K	A	n
400	0.065	0.43
420	0.014	0.55

Calculate enthalpies of adsorption for coverages $\theta = 0.1$ and $\theta = 0.5$, and comment on the results obtained.

21. Benzene was passed at various temperatures through a gas chromatography column packed with silica gel and the following specific retention volumes (extrapolated to zero sample size) were measured:

Temperature/°C	90	105	120	140
Specific retention volume/ (cm^3 at 0°C) g^{-1}	170	97	55	29

Calculate an isosteric enthalpy of adsorption of benzene on the silica gel sample.

22. Use the Kelvin equation to calculate the pore radius which corresponds to capillary condensation of nitrogen at 77 K and a relative pressure of 0.5. Allow for multilayer adsorption on the pore wall by taking the thickness of the adsorbed layer on a non-porous solid as 0.65 nm at this relative pressure. List the assumptions upon which this calculation is based. For nitrogen at 77 K, the surface tension is 8.85 mN m^{-1} and the molar volume is 34.7 cm^3 mol^{-1}.

23. At 20°C the surface tension of benzene is 28.9 mN m^{-1} and its molar volume is 89.2 cm^3 mol^{-1}. Determine the relative pressures at which condensation of benzene vapour should begin in a cylindrical capillary of radius 10 nm if the capillary is (*a*) closed at one end, (*b*) open at both ends. Assume zero contact angle and neglect adsorption on the walls of the capillary.

24. Iron has a body-centred cubic lattice (see Figure 5.16) with a unit cell side of 286 pm. Calculate the number of iron atoms per cm^2 of surface for each of the Fe(100), Fe(110) and Fe(111) crystal faces. Nitrogen adsorbs dissociatively on the Fe(100) surface and the LEED pattern is that of a C(2 × 2) adsorbed layer. Assuming saturation of this layer, calculate the number of adsorbed nitrogen atoms per cm^2 of surface.

25. The contact angle for water on paraffin wax is 105° at 20°C. Calculate the work of adhesion and the spreading coefficient. The surface tension of water at 20°C is 72.75 mN m^{-1}.

26. The pressure required to prevent liquid from entering a plug of a finely divided solid is twice as great for a liquid of surface tension 50 mN m^{-1}, which completely wets the solid, as it is for a liquid of surface tension 70 mN m^{-1}, which has a finite contact angle with the solid. Calculate this contact angle.

27. The following data refer to the adsorption of dodecanol from solution in toluene by a sample of carbon black, the BET nitrogen adsorption area of which is 105 m^2 g^{-1}:

Equilibrium concentration/ mol dm^{-3}	0.012	0.035	0.062	0.105	0.148
Amount adsorbed/ μmol g^{-1}	24.1	50.4	69.8	81.6	90.7

Show that the data fit a Langmuir adsorption isotherm equation and calculate the area occupied by each adsorbed dodecanol molecule at limiting adsorption.

28. Calculate the 'thickness' of the diffuse electric double layer for a negatively charged solid surface in contact with the following

aqueous solutions at 25°C: (*a*) 0.1 mol dm^{-3} KCl, (*b*) 0.001 mol dm^{-3} KCl, (*c*) 0.001 mol dm^{-3} K_2SO_4, (*d*) 0.001 mol dm^{-3} $MgCl_2$.

29. Electrokinetic measurements at 25°C on silver iodide in 10^{-3} mol dm^{-3} aqueous potassium nitrate give $d\zeta/d(pAg) = -35$ mV at the zero point of charge. Assuming no specific adsorption of K^+ or NO_3^- ions and no potential drop within the solid, estimate the capacity of the inner part of the electric double layer. Taking the thickness of the inner part of the double layer to be 0.4 nm, what value for the dielectric constant near to the interface does this imply? Comment on the result.

30. In a microelectrophoresis experiment a spherical particle of diameter 0.5 μm dispersed in a 0.1 mol. dm,$^{-3}$ aqueous solution of KCl at 25°C takes 8.0 s to cover a distance of 120 μm along one of the 'stationary levels' of the cell, the potential gradient being 10.0 V cm^{-1}. Calculate

(*a*) the electrophoretic mobility of the particle;
(*b*) the probable error in this single mobility determination arising from the Brownian motion of the particle during the course of the measurement;
(*c*) an approximate value for the zeta potential of the particle;
(*d*) an approximate value for the charge density at the surface of shear.

31. Spherical particles of radius 0.3 μm suspended in 0.02 mol dm^{-3} KCl(aq.) are observed to have an electrophoretic mobility of 4.0×10^{-8} m^2 s^{-1} V^{-1} at 25°C. Calculate an approximate value for the zeta potential. Briefly mention any simplifications upon which your calculation is based and in which sense they will affect your answer.

32. Calculate the rate of electro-osmotic flow of water at 25°C through a glass capillary tube 10 cm long and 1 mm diameter when the potential difference between the ends is 200 V. The zeta potential for the glass–water interface is −40 mV.

33. Spherical colloidal particles of diameter 10^{-7} m are dispersed in 10^{-2} mol dm^{-3} aqueous 1–1 electrolyte at 25°C. The Hamaker

constants for the particles and the dispersion medium are 1.6×10^{-19} J and 0.4×10^{-19} J, respectively, and the zeta potential is -40 mV. Using equations (8.7) and (8.10) to calculate V_R and V_A, respectively, calculate the total interaction energy between two of these particles when their shortest distance of approach, H, is 0.5 nm, 2 nm, 5 nm, 10 nm and 20 nm.

34. The following results were obtained by particle counting during the coagulation of a hydrosol at 25°C by excess 1–1 electrolyte:

Time/min	0	2	4	7	12	20
Particle concentration/10^8 cm^{-3}	100	14	8.2	4.6	2.8	1.7

Calculate a 'second-order' rate constant, k_2, and compare it with the value, k_2^0, calculated on the assumption that coagulation is a diffusion-controlled process.

35. The flow times in an Ostwald viscometer for solutions of polystyrene in toluene at 25°C are as follows:

Concentration/(g/100 cm^3)	0	0.4	0.8	1.2
Flow time/s	31.7	38.3	45.0	51.9

K and α in the expression, $[\eta] = KM_r^{\alpha}$, are 3.7×10^{-5} m^3 kg^{-1} and 0.62, respectively, for this polymer–solvent system. Assuming a constant density for the solutions, calculate an average relative molecular mass for the polystyrene sample. How would this relative molecular mass be expected to compare with the relative molecular mass of the same sample of polystyrene in toluene determined from (*a*) osmotic pressure and (*b*) light-scattering measurements?

36. The following viscosities were measured for solutions of cellulose acetate in acetone of concentration 0.5 g/100 cm^3:

$10^{-3}\ M_r$	85	138	204	302
$\eta/10^{-4}$ Pa s	5.45	6.51	7.73	9.40

The viscosity of acetone at the temperature of these measurements is 3.2×10^{-4} Pa s. Derive an expression from these data which could be used for routine determinations of the relative molecular masses of cellulose acetate samples in this relative molecular mass range from viscosity measurements. What information is given by this expression concerning the configuration of cellulose acetate molecules when dissolved in acetone?

Answers

1. 17.2 μm
2. M_r = 17 000 (i.e. M = 17.0 kg mol^{-1})
   f/f_0 = 1.11 (which, allowing for hydration, suggests that dissolved myoglobin molecules are approximately spherical – see Figure 2.1)
3. M_r = 34 000 (i.e. M = 34.0 kg mol^{-1}) (monodispersed)
4. M_r = 143 000 (i.e. M = 143 kg mol^{-1}) (number-average)
5. 0.2
6. M_r = 850 000 (i.e. M = 850 kg mol^{-1}) (mass-average)
7. 0.4 K of superheating
8. c.m.c. = 9×10^{-4} mol dm^{-3}
   A (at c.m.c.) = 50×10^{-20} m^2 $molecule^{-1}$
9.

c/mol dm^{-3}	0.01	0.02	0.04	0.08
π/mN m^{-1}	8.0	12.6	18.3	26.6
$\Gamma/10^{-6}$ mol m^{-2}	2.42	3.08	4.02	5.58
$A/10^{-20}$ m^2 $molecule^{-1}$	68.5	53.9	41.3	29.7

   For an ideal gaseous film at 293 K, $\pi A = kT = 4.04 \times 10^{-21}$ J.

10. c.m.c. = 7.8 mmol dm^{-3}
    A (at c.m.c.) = 39×10^{-20} m^2 $molecule^{-1}$
11. $\Delta H_{\text{micellisation}} = -6.2$ kJ mol^{-1}
12. (*a*) 43.8 mJ m^{-2}
    (*b*) (*i*) 43.6 mJ m^{-2}; (*ii*) 145.6 mJ m^{-2}
    (*c*) + 0.2 mJ m^{-2}
    (*d*) 22.0 mN m^{-1}

13. M_r = 13 000 (suggesting surface dissociation)

14. $\pi A = 2RT$ (suggesting that hydrogen is dissociated into atoms when chemisorbed on a nickel surface)

15. V_m = 225 cm^3 (s.t.p.) g^{-1}
a = 7.1 × 10^{-5} Pa^{-1}

16. 530 m^2 g^{-1}

17. 47 × 10^{-20} m^2

18. 7.8 m^2 g^{-1} (using BET equation)

19.

V/V_m	0.4	0.8	1.2
$\Delta H_{ads.}$/kJ mol^{-1}	−11.8	−12.7	−9.1

Values reflect multilayer physical adsorption on a fairly uniform solid surface

20. $\theta = 0.1$, $\Delta H_{ads.} = -180$ kJ mol^{-1}
$\theta = 0.5$, $\Delta H_{ads.} = -123$ kJ mol^{-1}

Absolute magnitudes suggest chemisorption and relative magnitudes reflect preferential adsorption at the more active sites on the solid surface

21. $\Delta H_{ads.} = -44$ kJ mol^{-1}

22. r = 2.03 nm (assuming zero contact angle, closed cylindrical pores, constancy of γ with r, equivalence of multilayer adsorption at flat and curved surfaces)

23. (*a*) $p/p_0 = 0.81$; (*b*) $p/p_0 = 0.90$

24. Fe(100), 12.2 × 10^{14} atoms cm^{-2};
Fe(110), 17.3 × 10^{14} atoms cm^{-2};
Fe(111), 7.1 × 10^{14} atoms cm^{-2};
Fe(100)–C(2 × 2)–N, 6.1 × 10^{14} N atoms cm^{-2}

25. W_a = 54.0 mJ m^{-2}, S = −91.5 mJ m^{-2}

26. 69°

27. 145 × 10^{-20} m^2

28. (*a*) 1 nm; (*b*) 10 nm; (*c*) 7 nm; (*d*) 5 nm

29. C (inner part of double layer) = 0.106 F m^{-2}
ϵ/ϵ_0 (inner part of double layer) = 4.8 (suggesting orientation of water molecules close to the surface)

30. (*a*) $u_E = 1.5 \times 10^{-8}$ m^2 s^{-1} V^{-1}
(*b*) 3.3 per cent
(*c*) 19.2 mV (using Smoluchowski equation; $\kappa a \approx 260$)
(*d*) 0.014 C m^{-2}

31. ζ = 51 mV (using Smoluchowski equation; $\kappa a \approx 140$; therefore, calculated ζ is probably an underestimate)

32. 5×10^{-5} cm^3 s^{-1} towards $-ve$ electrode

33.

H/nm	0.5	2	5	10	20
$V/10^{-21}$ J	−73	+62	+12	−10	−8

34. $k_2 = 4.8 \times 10^{-12}$ cm^3 s^{-1}
$k_2^0 = 4kT/3\eta = 6.2 \times 10^{-12}$ cm^3 s^{-1}

35. M_r = 118 000
M_r (light-scattering, mass-average) > M_r (viscosity) > M_r (osmotic pressure, number-average)

36. $[\eta] = 1.6 \times 10^{-5}\, M_r^{0.80}$ m^3 kg^{-1}
Average polymer configuration intermediate between extended and random

References

General

1. IUPAC, 'Manual of Symbols and Terminology in Colloid and Surface Chemistry', *Pure Appl. Chem.*, **31**, 578 (1972)
2. ADAMSON, A.W., *Physical Chemistry of Surfaces* (5th Ed.), Wiley-Interscience (1990)
3. AVEYARD, R. and HAYDON, D.A., *An Introduction to the Principles of Surface Chemistry*, Cambridge University Press (1973)
4. CADENHEAD, D.A. and DANIELLI, J.F. (editors), *Progress in Surface and Membrane Science*, Vols 1– , Academic Press (1964–)
5. HIEMENZ, P.C., *Principles of Colloid and Surface Chemistry* (2nd Ed.), Dekker (1986)
6. HIEMENZ, P.C., *Polymer Chemistry – The Basic Concepts*, Dekker (1984)
7. HUNTER, R.J., *Foundations of Colloid Science*, Oxford University Press, Volume 1 (1987), Volume 2 (1989)
8. KRUYT, H.R. (editor), *Colloid Science*, Elsevier: Volume 1, 'Irreversible Systems' (1952); Volume 2, 'Reversible Systems' (1949)
9. MATIJEVIC, E. (editor), *Surface and Colloid Science*, Vols 1– , Wiley-Interscience (1969–)
10. MORRISON, S.R., *The Chemical Physics of Surfaces* (2nd Ed.), Plenum (1990)
11. ROSEN, M.J., *Surfactants and Interfacial Phenomena*, Wiley (1978)
12. ROSS, S. and MORRISON, I.D., *Colloidal Systems and Interfaces*, Wiley (1988)
13. SOMORJAI, G.A., *Chemistry in Two Dimensions–Surfaces*, Cornell University Press (1981)
14. TADROS, Th.F. (editor), *Surfactants*, Academic Press (1984)
15. VAN OLPHEN, H. and MYSELS, K.J. (editors), *Physical Chemistry: Enriching Topics from Colloid and Surface Science*, IUPAC Commission 1.6, Theorex (1975)
16. VOLD, R.D. and VOLD, M.J., *Colloid and Interface Chemistry*, Addison-Wesley (1983)
17. WEISSBERGER, A. and ROSSITER, B.W. (editors), *Physical Methods of Chemistry*, Vol. 1, Parts 1–6 of *Techniques of Chemistry*, Wiley-Interscience (1971–77)

Text

Books and Review Articles

18. VAN OLPHEN, H., *An Introduction to Clay Colloid Chemistry* (2nd Ed.), Wiley (1977)

19. MATIJEVIC, E., 'Preparation and characterisation of monodispersed metal hydrous oxide sols', *Progr. Colloid Polymer Sci.*, **61**, 24–35 (1976); 'Monodispersed Metal (Hydrous) Oxides', *Acc. Chem. Res.*, **14**, 22–29 (1981)
20. ILER, R.K., 'Colloidal silica', in reference 9: **6**, 1–100 (1973)
21. HARKINS, W.D., *The Physical Chemistry of Surface Films*, Reinhold (1952)
22. HEARN, J., WILKINSON, M.C. and GOODALL, A.R., *Adv. Colloid Interface Sci.*, **14**, 173 (1981)
23. VAN DEN ESKER, M.W.J. and PIEPER, J.H.A., 'Preparation and critical coagulation concentration of a polystyrene latex', in reference 14, pp. 219–228
24. BUSCALL, R., CORNER, T. and STAGEMAN, J.F. (editors), *Polymer Colloids*, Elsevier (1985)
25. GILL, W.N., DERZANSKY, L.J. and DOSHI, M.R., 'Convective diffusion in laminar and turbulent hyperfiltration (reverse osmosis) systems', in reference 9: **4**, 261–360 (1971)
26. ONCLEY, J.L., in COHN, E.J. and EDSALL, J.T., *Proteins, Amino Acids and Peptides*, A.C.S. Monograph 90, Reinhold (1943)
27. DUNLOP, P.J., STEEL, B.J. and LANE, J.E., 'Experimental methods for studying diffusion in liquids, gases and solids', in reference 17, Part 4, pp. 205–349
28. ALEXANDER, A.E. and JOHNSON, P., *Colloid Science*, Oxford University Press (1949)
29. ALLEN, T., *Particle Size Measurement*, Chapman and Hall (1968)
30. SCHACHMANN, H.K., *Ultracentrifugation in Biochemistry*, Academic Press (1959)
31. KEGELES, G., YPHANTIS, D.A. and SCHNEIDER, R.F., 'Determination with the ultracentrifuge', in reference 17, Part 6, pp. 1–61
32. PICKELS, E.G., *Chem. Rev.*, **30**, 351 (1942)
33. OVERTON, J.R., 'Determination of osmotic pressure', in reference 17, Part 5, pp. 309–346
34. TOMBS, M.P. and PEACOCKE, A.R., *The Osmotic Pressure of Biological Macromolecules*, Clarendon Press (1974)
35. PETERLIN, A., 'Streaming and stress birefringence', in reference 120: **1**, 615–651 (1956)
36. SCOTT, R.G. and McKEE, A.N., 'Electron microscopy', in reference 15, Part 3B, pp. 291–438
37. GOODHEW, P.J., *Electron Microscopy and Analysis*, Wykeham (1975)
38. KERKER, M., *The Scattering of Light and Other Electromagnetic Radiation*, Academic Press (1969)
39. OSTER, G., 'Light scattering', in reference 17, Part 3B, pp. 75–111
40. BOHREN, C.E. and HUFFMAN, D.R., *Absorption and Scattering of Light by Small Particles*, Wiley (1983)
41. DE BOER, J.H., *The Dynamical Character of Adsorption* (2nd Ed.), Oxford University Press (1968)
42. FOWKES, F.M., 'Dispersion force contributions to surface and interfacial tensions, contact angles and heats of immersion', A.C.S. *Advances in Chemistry Series*, **43**, 99–111 (1964)
43. PADDAY, J.F., 'Measurement of surface tension', in reference 9: **1**, 39–251 (1969)
44. ALEXANDER, A.E., and HAYTER, J.B., 'Determination of surface and interfacial tension', in reference 17, Part 5, pp. 501–555
45. AMBWANI, D.S. and FORT, T., 'Pendant drop technique for measuring liquid boundary tensions', in reference 9: **11**, 93–119 (1979)
46. DEFAY, R. and PETRE, G., 'Dynamic surface tension', in reference 9: **3**, 27–82 (1971)

47. MURAMATSU, M., 'Radioactive tracers in surface and colloid science', in reference 9: **6**, 101–184 (1973)
48. HALL, D.G. and TIDDY, G.J.T., 'Surfactant solutions: dilute and concentrated', in LUCASSEN-REYNDERS, E.H. (editor), *Anionic Surfactants*, **11**, 55–108, Dekker (1981)
49. TANFORD, C., *The Hydrophobic Effect – Formation of Micelles and Biological Membranes*, Wiley (1973)
50. MUKERJEE, P., 'The nature of the association equilibria and hydrophobic bonding in aqueous solutions of association colloids', *Advan. Colloid Interface Sci.*, **1**, 241–275 (1967)
51. ELWORTHY, P.H., FLORENCE, A.T. and MACFARLANE, C.B., *Solubilisation by Surface Active Agents*, Chapman and Hall (1968)
52. CORDES, E. (editor), *Reaction Kinetics in Micelles*, Plenum (1973)
53. HALL, D.G. and PETHICA, B.A., 'Thermodynamics of micelle formation', in SHICK, M.J. (editor), *Non-Ionic Surfactants*, pp. 516–577, Arnold (1967)
54. DAVIES, J.T. and RIDEAL, E.K., *Interfacial Phenomena* (2nd Ed.), Academic Press (1963)
55. ADAM, N.K., *The Physics and Chemistry of Surfaces* (3rd Ed.), Oxford University Press (1941); Dover (1968)
56. GAINES, G.L., *Insoluble Monolayers at Liquid–Gas Interfaces*, Interscience (1966); ALEXANDER, A.E. and HIBBERD, G.E., 'Determination of properties of insoluble monolayers at mobile interfaces', in reference 17, Part 5, pp. 557–589
57. MACRITCHIE, F., *Chemistry of Interfaces*, Academic Press (1990)
58. JOLY, M., 'Rheological properties of monomolecular films', in reference 9: **5**, 1–193 (1972)
59. CHAPMAN, D. (editor), *Biological Membranes*, Academic Press (1968)
60. PRINCE, L.M. and SEARS, D.F. (editors), *Biological Horizons of Surface Science*, Academic Press (1973)
61. TIEN, H.T., 'Bimolecular lipid membranes', in reference 9: **4**, 361–423 (1971)
62. SINGER, S.J. and NICHOLSON, G.L., 'The fluid mosaic model of the structure of cell membranes', *Science*, **175**, 720–731 (1972)
63. GREGG, S.J. and SING, K.S.W., *Adsorption, Surface Area and Porosity* (2nd Ed.), Academic Press (1982)
64. PIEROTTI, R.A. and THOMAS, H.E., 'Physical adsorption: the interaction of gases with solids', in reference 9: **4**, 93–259 (1971)
65. CAMPBELL, I.M., *Catalysis at Surfaces*, Chapman and Hall (1988)
66. GASSER, R.P.H., *An Introduction to Chemisorption and Catalysis by Metals*, Oxford University Press (1985)
67. ROBERTS, M.W. and McKEE, C.S., *Chemistry of the Metal-Gas Interface*, Oxford University Press (1978)
68. TOMPKINS, F.C., *Chemisorption of Gases on Metals*, Academic Press (1978)
69. DYER, A., *Zeolite Molecular Sieves*, Wiley (1988)
70. WINSLOW, D.N., 'Advances in experimental techniques for mercury intrusion porosimetry', in reference 9: **13**, 259–286 (1984)
71. DRAIN, L.E., 'Thermodynamics of physical adsorption', *Sci. Progr.*, **42**, 608–628 (1954)
72. McCLELLAN, A.L. and HARNSBERGER, H.F., 'Cross-sectional areas of molecules adsorbed on solid surfaces', *J. Colloid Interface Sci.*, **23**, 577–599 (1967)
73. ROBERTS, M.W., *Chem. Soc. Rev.*, **6**, 373 (1977)
74. ERTL, G., 'Kinetics of chemical processes on well-defined surfaces', in ANDERSON, J.R. and BOUDART, M. (editors), *Catalysis-Science and Technology*, **4**, 209–282, Springer-Verlag (1983)

75. JOHNSON, R.E., and DETTRE, R., 'Wettability and contact angle', in reference 9: **2**, 85–153 (1969)
76. GOOD, R.J., 'Contact angles and the surface free energy of solids', in reference 9: **11**, 1–30 (1979)
77. FINCH, J.A. and SMITH, G.W., 'Contact angles and wetting', in LUCASSEN-REYNDERS, E.H. (editor), *Anionic Surfactants*, **11**, 317–384, Dekker (1981)
78. ZISMAN, W.A., 'Relation of the equilibrium contact angle to liquid and solid constitution', A.C.S. *Advances in Chemistry Series*, **43**, 1–51 (1964)
79. NEUMANN, A.W. and GOOD, R.J., 'Techniques of measuring contact angles', in reference 9: **11**, 31–92 (1979)
80. HORNSBY, D. and LEJA, J., 'Selective flotation and its surface chemical characteristics', in reference 9: **12**, 217–313 (1982)
81. IVES, K.J. (editor), *The Scientific Basis of Flotation*, Nijhoff (1984)
82. SCHWARTZ, A.M., 'The physical chemistry of detergency', in reference 9: **5**, 195–244 (1972)
83. CUTLER, W.G. and DAVIES, R.C., *Detergency: Theory and Test Methods*, Dekker, Part 1 (1972); Part 2 (1975); Part 3 (1981)
84. SCHWARTZ, A.M., 'Research techniques in detergency', in reference 9: **11**, 305–334 (1979)
85. KUSHNER, L.M. and HOFFMAN, J.I., 'Synthetic detergents', *Sci. Am.* (October 1951), pp. 26–30
86. SCHAY, G., 'Adsorption of solutions of nonelectrolytes', in reference 9: **2**, 155–211 (1969); 'Thermodynamics of adsorption from solution', in reference 15: pp. 229–250
87. OVERBEEK, J.Th.G., 'Electrochemistry of the double layer', in reference 8, Vol. 1, pp. 115–193
88. LOEB, A.L., OVERBEEK, J.Th.G. and WIERSEMA, P.H., *The Electrical Double Layer around a Spherical Particle*, M.I.T. Press (1961)
89. GRAHAME, D.C., 'The electrical double layer and the theory of electrocapillarity', *Chem. Rev.*, **41**, 441–501 (1947)
90. OVERBEEK, J.Th.G., 'Electrokinetic phenomena', in reference 8, Vol. 1, pp. 194–244
91. SMITH, A.L., 'Electrical phenomena associated with the solid–liquid interface', in PARFITT, G.D. (editor), *Dispersions of Powders in Liquids* (2nd Ed.), pp. 86–131, Applied Science (1973)
92. HUNTER, R.J., *Zeta Potential in Colloid Science*, Academic Press (1981)
93. SHAW, D.J., *Electrophoresis*, Academic Press (1969)
94. SMITH, I., *Zone Electrophoresis*, Heinemann (1976); ANDREWS, A.T., *Electrophoresis: Theory, Techniques and Biochemical Applications*, Oxford University Press (1981)
95. ISRAELACHVILI, J.N., *Intermolecular and Surface Forces with Applications to Colloidal and Biological Systems*, Academic Press (1985)
96. OVERBEEK, J.Th.G., 'Stability of hydrophobic colloids and emulsions', in reference 8, Vol. 1, pp. 302–341
97. VERWEY, E.J.W. and OVERBEEK, J.Th.G., *Theory of the Stability of Lyophobic Colloids*, Elsevier (1948)
98. CLUNIE, J.S., GOODMAN, J.F. and INGRAM, B.T., 'Thin liquid films', in reference 9: **3**, 167–240 (1971)
99. OVERBEEK, J.Th.G., 'Recent developments in the understanding of colloid stability', *J. Colloid Interface Sci.*, **58**, 408–422 (1977)
100. GREGORY, J., 'Flocculation by inorganic salts', in IVES, K.J. (editor), *The Scientific Basis of Flocculation*, pp. 89–99, N.A.T.O. Advanced Studies Institute Series, Sijthoff and Noordhoff (1975)

101. MATIJEVIC, E., 'The role of chemical complexing in the formation and stability of colloidal dispersions', *J. Colloid Interface Sci.*, **58**, 374–389 (1977)
102. MAHANTY, J.W. and NINHAM, B.W., *Dispersion Forces*, Academic Press (1975)
103. PARSEGIAN, V.A., 'Long range van der Waals forces', in reference 15, pp. 27–72
104. GREGORY, J., 'The calculation of Hamaker constants', *Advan. Colloid Interface Sci.*, **2**, 396–417 (1969)
105. VISSER, J., 'Adhesion of colloidal particles', in reference 9: **8**, 3–84 (1976)
106. OTTEWILL, R.H., 'Colloid chemistry – today and tomorrow', *Progr. Colloid Polymer Sci.*, **59**, 14–26 (1976); 'Stability and instability in dispersed systems', *J. Colloid Interface Sci.*, **58**, 357–373 (1977)
107. DERYAGIN, B.V., ABRICOSSOVA, I.I. and LIFSHITZ, E.M., 'Direct measurement of molecular attraction between solids separated by a narrow gap', *Q. Rev. Chem. Soc.*, **10**, 295–329 (1956); DERYAGIN, B.V., 'The force between molecules', *Sci. Am.* (July 1960), 3–9
108. ISRAELACHVILI, J.N. and TABOR, D., 'van der Waals forces: theory and experiment', in CADENHEAD, D.A. and DANIELLI, J.F. (editors), *Progress in Surface and Membrane Science*, **7**, 1–55, Academic Press (1973)
109. OVERBEEK, J.Th.G., 'The rule of Schulze and Hardy', *Pure Appl. Chem.*, **52**, 1151 (1980)
110. OVERBEEK, J.Th.G., 'Kinetics of flocculation', in reference 8, Vol. 1, pp. 278–301
111. GREGORY, J., 'Effects of polymers on colloid stability', in *Scientific Basis of Flocculation* (see reference 100), pp. 101–130
112. NAPPER, D.H., 'Steric stabilisation', *J. Colloid Interface Sci.*, **58**, 390–407 (1977)
113. SATO, T. and RUCH, R., *Stabilisation of Colloidal Dispersions by Polymer Adsorption*, Dekker (1980)
114. NAPPER, D.H., *Polymeric Stabilisation of Colloidal Dispersions*, Academic Press (1983)
115. PARFITT, G.D. and PEACOCK, J., 'Stability of colloidal dispersions in non-aqueous media', in reference 9, **10**, 163–266 (1978)
116. LYKLEMA, J., 'Principles of the stability of lyophobic colloidal dispersions in non-aqueous media', *Advan. Colloid Interface Sci.*, **2**, 65–114 (1968)
117. KITCHENER, J.A., 'Flocculation in mineral processing', in *Scientific Basis of Flocculation* (see reference 100), pp. 283–328
118. SHERMAN, P., *Industrial Rheology*, Academic Press (1970)
119. SCOTT-BLAIR, G.W., *An Introduction to Biorheology*, Elsevier (1974)
120. EIRICH, F.R. (editor), *Rheology – Theory and Applications*, Vols 1– , Academic Press (1956–)
121. JOHNSON, J.F., MARTIN, J.R. and PORTER, R.S., 'Determination of viscosity', in reference 17, Part 6, pp. 63–128
122. FERRY, J.D., *Viscoelastic Properties of Polymers* (3rd Ed.), Wiley (1980)
123. SHERMAN, P. (editor), *Emulsion Science*, Academic Press (1968)
124. LISSANT, K.J. (editor), *Emulsions and Emulsion Technology*, Dekker (1974)
125. BECHER, P. (editor), *Encyclopedia of Emulsion Technology*, Dekker (1981)
126. SHINODA, K. and FRIBERG, S., *Emulsions and Solubilisation*, Wiley (1986)
127. SHINODA, K. and FRIBERG, S., 'Microemulsions: colloidal aspects', *Advan. Colloid Interface Sci.*, **4**, 281–300 (1975)
128. PRINCE, L.M. (editor), *Microemulsions – Theory and Practice*, Academic Press (1977)
129. BIKERMANN, J.J., *Foams*, Springer-Verlag (1973)

130. SEBBA, F., *Foams and Biliquid Foams – Aprhons*, Wiley (1987)
131. REIST, P.C., *Introduction to Aerosol Science*, Macmillan (1984)

Research Papers

132. ZAISER, E.M. and LAMER, V.K., *J. Colloid Sci.*, **3**, 571 (1948)
133. OTTEWILL, R.H. AND WOODBRIDGE, R.F., *J. Colloid Sci.*, **16**, 581 (1961)
134. STROBER, W., FINK, A. and BOHN, E., *J. Colloid Interface Sci.*, **26**, 62 (1968)
135. OTTEWILL, R.H. and SHAW, J.N., *Kolloid Zh.*, **215**, 161 (1967)
136. ARCHIBALD, W.J., *J. Phys. Chem.*, **51**, 1204 (1947)
137. HUGGINS, M.L., *J. Chem. Phys.*, **9**, 440 (1941); *J. Phys. Chem.*, **46**, 151 (1942)
138. FLORY, P.J., *J. Chem. Phys.*, **9**, 660 (1941); **10**, 51 (1942); **13**, 453 (1945)
139. FUOSS, R.M. and MEAD, D.J., *J. Phys. Chem.*, **47**, 59 (1943)
140. DEBYE, P., *J. Phys. Chem.*, **51**, 18 (1947)
141. LAMER, V.K. and BARNES, M.D., *J. Colloid Sci.*, **1**, 71, 79 (1946)
142. ZIMM, B.H., *J. Chem. Phys.*, **16**, 1093, 1099 (1948)
143. ROSS, D.A., DHADWAL, H.S. and DYOTT, R.B., *J. Colloid Interface Sci.*, **64**, 533 (1978)
144. BOUCHER, E.A., GRINCHUK, T.M. and ZETTLEMOYER A.C., *J. Colloid Interface Sci.*, **23**, 600 (1967)
145. HARKINS, W.D. and JORDAN, H.F., *J. Am. Chem. Soc.*, **52**, 1751 (1930)
146. ZUIDEMA, H.H. and WATERS, G.W., *Ind. Eng. Chem. (Anal. Ed.)*, **13**, 312 (1941)
147. BOUCHER, E.A. and EVANS, M.J.B., *Proc. Roy. Soc.*, **A346**, 349 (1975)
148. HARKINS, W.D. and BROWN, F.E., *J. Am. Chem. Soc.*, **41**, 499 (1919)
149. LANDO, J.L. and OAKLEY, H.T., *J. Colloid Interface Sci.*, **25**, 526 (1967)
150. OWENS, D.K., *J. Colloid Interface Sci.*, **29**, 496 (1969)
151. HAYDON, D.A. and PHILLIPS, J.N., *Trans. Faraday Soc.*, **54**, 698 (1958); WEIL, L., *J. Phys. Chem.*, **70**, 133 (1966)
152. McBAIN, J.W. and SWAIN, R.C., *Proc. Roy. Soc.*, **A154**, 608 (1936)
153. CLIFFORD, J. and PETHICA, B.A., *Trans. Faraday Soc.*, **60**, 1483 (1964); **61**, 182 (1965)
154. CORKILL, J.M., GOODMAN, J.F. and WALKER, T., *Trans. Faraday Soc.*, **63**, 768 (1967); WALKER, T., *J. Colloid Interface Sci.*, **45**, 372 (1973)
155. ELWORTHY, P.H. and MYSELS, K.J., *J. Colloid Interface Sci.*, **21**, 331 (1966)
156. BROOKS, J.H. and PETHICA, B.A., *Trans. Faraday Soc.*, **60**, 209 (1964)
157. HUHNERFUSS, H., *J. Colloid Interface Sci.*, **107**, 84 (1985)
158. INOKUCHI, K., *Bull. Chem. Soc. Japan*, **27**, 203 (1954)
159. CUMPER, C.W.N. and ALEXANDER, A.E., *Trans. Faraday Soc.*, **46**, 235 (1950)
160. GORTER, E. and GRENDEL, F., *J. Exptl. Med.*, **41**, 439 (1925)
161. DANIELLI, J.F. and DAVSON, H., *J. Cell. Comp. Physiol.*, **5**, 495 (1934)
162. SCHULMAN, J.H. and RIDEAL, E.K., *Proc. Roy. Soc.*, **B122**, 46 (1937)
163. TITOFF, A., *Z. Phys. Chem.*, **74**, 641 (1910)
164. McBAIN, J.W. and BAKR, A.M., *J. Am. Chem. Soc.*, **48**, 690 (1926)
165. NELSON, F.M. and EGGERSTEN, F.T., *Anal. Chem.*, **30**, 1387 (1958)
166. AMBERG, C.H., SPENCER, W.B. and BEEBE, R.A., *Can. J. Chem.*, **33**, 305 (1955)
167. BRUNAUER, S., EMMETT, P.H. and TELLER, E., *J. Am. Chem. Soc.*, **60**, 309 (1938)
168. BRUNAUER, S., DEMING, L.S., DEMING, W.S. and TELLER, E., *J. Am. Chem. Soc.*, **62**, 1723 (1940)

169. BEECK, O., *Discuss. Faraday Soc.*, **7**, 118 (1950)
170. KARNAUKHOV, A.P., *J. Colloid Interface Sci.*, **103**, 311 (1985)
171. ERTL, G., PRIGGE, D., SCHLOEGL, R. and WEISS, M., *J. Catalysis*, **79**, 359 (1983)
172. ERTL, G. and THIELE, N., *Appl. Surface Sci.*, **3**, 99 (1979)
173. LANGMUIR, I. and SCHAEFFER, V., *J. Am. Chem. Soc.*, **59**, 2405 (1937); FORT, T. and PATTERSON, H.T., *J. Colloid Sci.*, **18**, 217 (1963)
174. JONES, W.C. and PORTER, M.C., *J. Colloid Interface Sci.*, **24**, 1 (1967)
175. BARTELL, F.E. *et al.*, *J. Phys. Chem.*, **36**, 3115 (1932); **38**, 503 (1934)
176. BRUIL, H.G. and VAN AARSTEN, J.J., *Colloid Polym. Sci.*, **252**, 32 (1974)
177. INNES, W.B. and ROWLEY, H.H., *J. Phys. Chem.*, **51**, 1176 (1951)
178. BLACKBURN, A., KIPLING, J.J. and TESTER, D.A., *J. Chem. Soc.*, **2373** (1957)
179. EIRICH, F.R., *J. Colloid Interface Sci.*, **58**, 423 (1977)
180. YOON, R.H. and YORDAN, J.L., *J. Colloid Interface Sci.*, **113**, 430 (1986)
181. LYKLEMA, J., *Kolloid Zh.*, **175**, 129 (1961); *Discuss. Faraday Soc.*, **42**, 81 (1966)
182. OTTEWILL, R.H. and WOODBRIDGE, R.F., *J. Colloid Sci.*, **19**, 606 (1964)
183. LEVINE, S. and BELL, G.M., *J. Colloid Sci.*, **17**, 838 (1962); LEVINE, S., MINGINS, J. and BELL, G.M., *J. Electroanal. Chem.*, **13**, 280 (1967)
184. TISELIUS, A., *Trans. Faraday Soc.*, **33**, 524 (1937)
185. OTTEWILL, R.H. and SHAW, J.N., *Kolloid Zh.*, **218**, 34 (1967)
186. LONG, R.P. and ROSS, S., *J. Colloid Interface Sci.*, **26**, 434 (1968)
187. HENRY, D.C., *Proc. Roy. Soc.*, **A133**, 106 (1931)
188. BOOTH, F., *Trans. Faraday Soc.*, **44**, 955 (1948); HENRY, D.C., *Trans. Faraday Soc.*, **44**, 1021 (1948)
189. GHOSH, B.N. *et al.*, *J. Indian Chem. Soc.*, **32**, 31 (1955); **40**, 425 (1963)
190. WIERSEMA, P.H., LOEB, A.L. and OVERBEEK, J.Th.G., *J. Colloid Interface Sci.*, **22**, 78 (1966)
191. SHAW, J.N. and OTTEWILL, R.H., *Nature (Lond.)*, **208**, 681 (1965)
192. LYKLEMA, J. and OVERBEEK, J.Th.G., *J. Colloid Sci.*, **16**, 501 (1961)
193. STIGTER, D., *J. Phys. Chem.*, **68**, 3600 (1964); HUNTER, R.J., *J. Colloid Interface Sci.*, **22**, 231 (1966)
194. DERYAGIN, B.V. and LANDAU, L., *Acta Phys. Chim. URSS*, **14**, 633 (1941)
195. HOGG, R., HEALY, T.W. and FUERSTENAU, D.W., *Trans. Faraday Soc.*, **62**, 1638 (1966); WIESE, G.R. and HEALY, T.W., *Trans. Faraday Soc.*, **66**, 490 (1970)
196. REERINK, H. and OVERBEEK, J.Th.G., *Discuss. Faraday Soc.*, **18**, 74 (1954)
197. HAMAKER, H.C., *Physica*, **4**, 1058 (1937)
198. LIFSHITZ, E.M., *Sov. Phys.*, **2**, 73 (1956)
199. NINHAM, B.W. and PARSEGIAN, V.A., *Biophys. J.*, **10**, 646 (1970)
200. BARCLAY, L.M. and OTTEWILL, R.H., *Spec. Discuss. Faraday Soc.*, **1**, 138 (1970); CALLAGHAN, I.C. and OTTEWILL, R.H., *Faraday Discuss. Chem. Soc.*, **57**, 110 (1975)
201. ROBERTS, A.D. and TABOR, D., *Proc. Roy. Soc.*, **A325**, 323 (1971)
202. TABOR, D. and WINTERTON, R.H.S., *Proc. Roy. Soc.*, **A312**, 435 (1969)
203. ISRAELACHVILI, J.N. and TABOR, D., *Nature (Phys. Sci.)*, **236**, 106 (1972); *Proc. Roy. Soc.*, **A331**, 19 (1972)
204. OTTEWILL, R.H. and SHAW, J.N., *Discuss. Faraday Soc.*, **42**, 154 (1966)
205. LICHTENBELT, J.W.Th., RAS, H.J.M.C. and WIERSEMA, P.H., *J. Colloid Interface Sci.*, **46**, 522 (1974); LICHTENBELT, J.W.Th., PATHMAMANOHARAN, C. and WIERSEMA, P.H., *J. Colloid Interface Sci.*, **49**, 281 (1974)
206. FAIRHURST, D., Thesis, Liverpool Polytechnic (1969)

207. VOLD, M.J., *J. Colloid Sci.*, **16**, 1 (1961)
208. OSMOND, D.W.J., VINCENT, B. and WAITE, F.A., *J. Colloid Interface Sci.*, **42**, 262 (1973); VINCENT, B., *J. Colloid Interface Sci.*, **42**, 270 (1973)
209. LANGBEIN, D., *J. Adhesion*, **1**, 237 (1969)
210. MACKOR, E.L., *J. Colloid Sci.*, **6**, 492 (1951)
211. DOLAN, A.K. and EDWARDS, S.F., *Proc. Roy. Soc.*, **A337**, 509 (1974); **A343**, 427 (1975)
212. HESSELINK, F.Th., VRIJ, A. and OVERBEEK, J.Th.G., *J. Phys. Chem.*, **75**, 65 2094 (1971)
213. CAIRNS, R.J.R. and OTTEWILL, R.H., *J. Colloid Interface Sci.*, **54**, 45 (1976); HOMOLA, A. and ROBERTSON, A.A., *J. Colloid Interface Sci.*, **54**, 286 (1976)
214. MATIJEVIC, E. and OTTEWILL, R.H., *J. Colloid Sci.*, **13**, 242 (1958)
215. HEALY, T.W. and LAMER, V.K., *J. Colloid Sci.*, **19**, 323 (1964)
216. SLATER, R.W. and KITCHENER, J.A., *Discuss. Faraday Soc.*, **42**, 267 (1966)
217. VOET, A., *J. Phys. Chem.*, **61**, 301 (1957)
218. BOYD, J., PARKINSON, C. and SHERMAN, P., *J. Colloid Interface Sci.*, **41**, 359 (1972)
219. SCHULMAN, J.H. and COCKBAIN, E.G., *Trans. Faraday Soc.*, **36**, 651 (1940)
220. VOLD, R.D. and MITTAL, K.L., *J. Colloid Interface Sci.*, **38**, 451 (1972)
221. VOLD, R.D. and GROOT, R.C., *J. Colloid Sci.*, **19**, 384 (1964)
222. GRIFFIN, W.C., *J. Soc. Cosmetic Chemists*, **1**, 311 (1949); **5**, 4 (1954)
223. SHINODA, K. and ARAI, H., *J. Phys. Chem.*, **68**, 3485 (1964)
224. SHINODA, K. and SAITO, H., *J. Colloid Interface Sci.*, **30**, 258 (1969)
225. DERYAGIN, B.V. and TITIJEVSKAYA, A.S., *Proc. Second Int. Congr. Surface Activity*, Vol. 1, p. 210, Butterworths (1957)
226. OVERBEEK, J.Th.G., *J. Phys. Chem.*, **64**, 1178 (1960); MYSELS, K.J., *J. Phys. Chem.*, **68**, 3441 (1964)
227. VAN DEN TEMPEL, M., *J. Colloid Sci.*, **13**, 125 (1958)
228. PRINS, A. and VAN DEN TEMPEL, M., *Spec. Discuss. Faraday Soc.*, **1**, 20 (1970)
229. EWERS, W.E. and SUTHERLAND, K.L., *Australian J. Sci. Res.*, **A5**, 697 (1952); SHEARER, L.T. and AKERS, W.W., *J. Phys. Chem.*, **62**, 1264 (1958)
230. ROSS, S. and HAAK, R.M., *J. Phys. Chem.*, **62**, 1260 (1958)

Index

图书在版编目（CIP）数据

胶体和表面化学导论 ： 第4版 ： 英文 / （英） 邓肯·J. 肖 (Duncan J. Shaw) 著. -- 影印本. -- 北京 : 世界图书出版有限公司北京分公司, 2025. 8. -- ISBN 978-7-5232-2344-4

Ⅰ. O648; O647.11

中国国家版本馆CIP数据核字第2025C7P058号

中文书名	胶体和表面化学导论（第 4 版）
英文书名	Introduction to Colloid and Surface Chemistry, 4th Edition
著　　者	[英] 邓肯 · J. 肖（Duncan J. Shaw）
责任编辑	陈　亮　刘叶青
出版发行	世界图书出版有限公司北京分公司
地　　址	北京市东城区朝内大街 137 号
邮　　编	100010
电　　话	010-64038355（发行）　64033507（总编室）
网　　址	http://www.wpcbj.com.cn
邮　　箱	wpcbjst@vip.163.com
销　　售	新华书店
印　　刷	北京建宏印刷有限公司
开　　本	710mm × 1000mm　1/16
印　　张	19.75
字　　数	310 千字
版　　次	2025 年 8 月第 1 版
印　　次	2025 年 8 月第 1 次印刷
版权登记	01–2022–3008
国际书号	ISBN 978-7-5232-2344-4
定　　价	79.00 元
